Cognitive Ergonomics:
Understanding, Learning and Designing Human–Computer Interaction

Computers and People Series

Edited by

B. R. GAINES and A. MONK

The series is concerned with all aspects of human–computer relationships, including interaction, interfacing modelling and artificial intelligence. Books are interdisciplinary, communicating results derived in one area of study to workers in another. Applied, experimental, theoretical and tutorial studies are included.

Monographs

On Becoming a Personal Scientist: Interactive computer elicitation of personal models of the world, *Mildred L. G. Shaw* 1980

Communicating with Microcomputers: An introduction to the technology of man–computer communication, *Ian H. Witten* 1980

The Computer in Experimental Psychology, *R. Bird* 1981

Recent Advances in Personal Construct Technology, *Mildred L. G. Shaw* 1981

Principles of Computer Speech, *I. H. Witten* 1982

Cognitive Psychology of Planning, *J-M. Hoc* 1988

Edited Works

Computing Skills and the User Interface, *M. J. Coombs and J. L. Alty (eds)* 1981

Fuzzy Reasoning and Its Applications, *E. H. Mamdani and B. R. Gaines (eds)* 1981

Intelligent Tutoring Systems, *D. Sleeman and J. S. Brown (eds)* 1982 (1986 paperback)

Designing for Human–Computer Communication, *M. E. Sime and M. J. Coombs (eds)* 1983

The Psychology of Computer Use, *T. R. G. Green, S. J. Payne and G. C. van der Veer (eds)* 1983

Fundamentals of Human–Computer Interaction, *Andrew Monk (ed)* 1984, 1985

Working with Computers: Theory versus Outcome, *G. C. van der Veer, T. R. G. Green, J-M. Hoc and D. Murray (eds)* 1988

Cognitive Engineering in Complex Dynamic Worlds, *E. Hollnagel, G. Mancini and D. D. Woods (eds)* 1988

Computers and Conversation, *P. Luff, N. Gilbert and D. Frohlich (eds)* 1990

EACE Publications

(Consulting Editors: *Y. WAERN and J-M. HOC*)

Cognitive Ergonomics, *P. Falzon (ed)* 1990

Psychology of Programming, *J-M. Hoc, T. R. G. Green, R. Samurçay and D. Gilmore (Forthcoming)*

Cognitive Ergonomics:
Understanding, Learning and Designing Human–Computer Interaction

Edited by

P. FALZON
INRIA, Rocquencourt, BP 105-78150, Le Chesnay, France

A Publication of the
European Association of Cognitive Ergonomics

ACADEMIC PRESS

Harcourt Brace Jovanovich, Publishers
London   San Diego   New York
Boston   Sydney   Tokyo   Toronto

ACADEMIC PRESS LTD.
24/28 Oval Road,
London NW1 7DX

United States Edition published by
ACADEMIC PRESS INC.
San Diego, California 92101-4311

Copyright ©1990 by
ACADEMIC PRESS LTD.

This book is printed on acid-free paper ∞

British Library Cataloguing in Publication Data

Is available

ISBN 0-12-248290-5

Filmset by Bath Typesetting Ltd., Bath, Avon
Printed in Great Britain by St Edmundsbury Press Ltd,
Bury St Edmunds, Suffolk

CONTENTS

SECTION 1
MODELS FOR DESIGN

v

SECTION 2
LEARNING PROCESSES

SECTION 3
PLANNING AND UNDERSTANDING

CONTRIBUTORS

ACKERMANN, D. Work and Organizational Psychology Unit, ETH Zürich, CH-8092 Zürich, Switzerland.

ALLWOOD, C. M. Department of Psychology, University of Göteborg, Box 14158, S 400-20 Göteborg, Sweden.

BARNARD, P. MRC Applied Psychology, Cambridge, U.K.

BISSERET, A. INRIA, Rocquencourt, BP 105-78150, Le Chesnay, France.

DÉTIENNE, F. INRIA, Rocquencourt, BP 105-78150, Le Chesnay, France.

FALZON, P. INRIA, Rocquencourt, BP 105-78150, Le Chesnay, France.

FELT, M. A. M. Department of Psychology, Free University, De Boelelaan 1111, 1081 HV Amsterdam, Netherlands

GOODSTEIN, L. P. Present address: Ibsgarden 196, DK 4000 Roskilde, Denmark.

GREUTMANN, T. ETH Zürich, Lehrstuhl für Arbeits- und Organisationspsychologie, Nelkenstr. 11, 8092 Zürich, Switzerland.

HOC, J-M. CNRS-Université de Paris 8, Psychologie Cognitive du Traitement de L'Information Symbolique, 2 Rue de la Liberté, F-93526 Saint-Denis, Cedex 2, France.

HOPPE, H. U. Fraunhofer-Institut für Arbeitswirtschaft und Organisation (IAO), Holzgartenstr. 17, 7000 Stuttgart 1, FRG.

LANSDALE, M. Cognitive Ergonomics Research Group, Department of Human Sciences, University of Technology, Loughborough, Leicestershire, U.K.

MACLEAN, A. Rank Xerox Ltd, Euro PARC, Cambridge, U.K.

SAMURÇAY, R. CNRS-Université de Paris 8, URA 1297 'Psychologie Cognitive du Traitement de l'Information Symbolique', Equipe de Psychologie Cognitive Ergonomique, 2, Rue de la Liberté, F-93526 Saint-Denis, Cedex 2, France.

SCHINDLER, R. Department of Psychology, Humboldt University, Berlin, GDR.

SCHUSTER, A. Department of Psychology, Humboldt University, Berlin, GDR.

STELOVSKY, J. Department of Computer Science, University of Honolulu, Hawaii 94822, U.S.A.

VAN DER VEER, G. C. Department of Psychology, Free University, De Boelelaan 1111, 1081 HV Amsterdam, Netherlands

VOSSEN, P. H. Fraunhofer-Institut für Arbeitswirtschaft und Organisation (IAO), Holzgartenstr. 17, 7000 Stuttgart 1, FRG.

WAERN, Y. Department of Psychology, University of Stockholm, S106-91 Stockholm, Sweden.

WHITEFIELD, A. Ergonomics Unit, University College London, 26 Bedford Way, London WC1H 0AP, U.K.

WIJK, R. Department of Psychology, Free University, De Boelelaan 1111, 1081 HV Amsterdam, Netherlands

WILSON, M. Rutherford Appleton Laboratory, Chilton, Oxfordshire, U.K.

ZIEGLER, J. E. Fraunhofer-Institut für Arbeitswirtschaft und Organisation (IAO), Holzgartenstr. 17, 7000 Stuttgart 1, FRG.

PREFACE

In the fall of 1986, the Third European Conference on Cognitive Ergonomics (ECCE-3) was held in Paris. Sixty people attended, from all over Europe: Denmark, Finland, France, Germany, Great Britain, Italy, the Netherlands, Sweden, Switzerland (plus one attendee from ... Canada). During the conference, the decision to lay the foundations of the European Association of Cognitive Ergonomics (EACE) took place.

This book gathers selected papers submitted at the conference, which have been revised for this publication. The book is divided into three sections: models for design, learning processes, planning and understanding. Each section begins with an introduction, which puts the papers of the section into perspective, so I do not think it useful to provide an introduction to the introductions. Instead, I like to emphasize the coverage of the chapters in this book, starting from a brief reminder of what cognitive ergonomics is about.

Cognitive ergonomics can be defined as the subfield of cognitive science especially concerned with human task-oriented activity. A book on the cognitive ergonomics of human–computer interaction can thus be viewed from two standpoints: either from the standpoint of the tasks being performed (or application domains dealt with), or from the standpoint of the cognitive activities involved in completing the tasks.

Two lists will help to appreciate the extension of the coverage. Concerning tasks and application domains, the reader will find studies of advisory interaction, consultation dialogues, interactions with a data base, dialogue grammars, assessment techniques, calculation using a pocket calculator, spreadsheet calculation, font edition, graphic edition, word processing, text design, mechanical design, interface design, management of filing systems, office organization, process control, assistance systems, program understanding, programming, video games.

The following research areas are illustrated: action regulation, automatization of behaviour, analogical thinking, generalization, inferences, metaphors, dialogue strategies, human–computer and human–human dialogues, natural language, interaction models, interaction modes, metacommunication, knowledge organization, schema theory, mental models, metaknowledge, memory, information retrieval, knowledge transfer, transfer of skill, learning, restructuring, learning methods, training, planning, understanding, decision making, design.

Two facts are worth noticing. First, this variety of themes and domains should not be interpreted as a sign of dispersion. On the contrary, it is striking to notice how much the authors have in common both on a theoretical level and in their domains of interest. This convergence evident in the section introductions, which do not need to twist the texts in order to make them fit into a single framework. Thus, the chapters provide a non-discordant view of what European research in cognitive ergonomics is concerned with.

The second fact is that research appears well integrated not only within Europe, but also with research outside Europe. References to non-European works abound in every text, showing that European research is not confined to its geographic boundaries.

The book, and the creation of EACE, give evidence of the strength and maturity of European research in human–computer interaction.

Pierre Falzon

SECTION 1
MODELS FOR DESIGN

MODELS FOR DESIGN: AN INTRODUCTION

Pierre FALZON

INRIA, Rocquencourt, BP 105, 78150, Le Chesnay, France

The following chapters in this Section consider the use of models for the design of interactive software. The necessity of models is widely acknowledged, and many have been proposed in the past decade. The initial part of Whitefield's chapter is devoted to a critical overview of human–computer interaction models from the standpoint of their roles in design. Whitefield attempts to identify the entities that may be modelled, and the actors that can elaborate models of these entities. Whitefield identifies five entities: the researcher, the designer, the program, the user and the system. The system is taken in its ergonomic sense, i.e. the combination of the user and the program. These entities can be modelled by four actors: the researcher, the designer, the progam and the user. The interaction between actors and entities allows one to define classes of models in which existing theories can be categorized. A model can be evaluated on its predictive validity: what is the fit of the model with subjects' actual performance? And secondly it can be evaluated on its usefulness and usability: what is the interest of the model for the designer's activity? Is the model easy to use?

The study of Ziegler, Vossen and Hoppe illustrates the first type of evaluation. These authors assess the predictive value of Polson and Kieras's model of cognitive behaviour, cognitive complexity theory (CCT: Polson and Kieras, 1985). CCT is derived from the GOMS model (Card et al., 1983), which allows description of the user's task as a tree of goals, operators, methods and selection rules. CCT can transform the GOMS analysis of a given task in production rules. Thus, a CCT model will allow the simulation of the system behaviour, and the prediction of some parameters such as execution time, learning time (which varies according to the number of new productions to be acquired), and transfer of skill (the number of productions which are common to the known task and the new one).

Ziegler, Vossen and Hoppe assess the validity of CCT using text and graphics editing tasks. The same commands can be applied in both domains

COGNITIVE ERGONOMICS:
UNDERSTANDING, LEARNING AND DESIGNING
HUMAN–COMPUTER INTERACTION

(functional equivalence). The tasks (text and graphics editing) are very similar, i.e. they share many production rules; CCT thus predicts a strong transfer between the two tasks. The actual performance of subjects on these tasks is compared with the predictions made by CCT, and this comparison validates CCT as a predictive model. The observed differences (between performance and predictions) can be attributed to differences between the training procedure in Polson and Kieras's study and that used in this experiment. The authors conclude that, although CCT presents several limitations (lack of prediction of interindividual variations, consideration of a unique strategy, simplicity of the tasks), it is effective in predicting user behaviour.

The second part of Whitefield's chapter addresses the second type of evaluation described above (model usefulness and usability), and focuses on the analysis of two models: KLM, the Keystroke-Level Model of Card et al. (1983), and BDM, the Blackboard Design Model (Whitefield, 1989). After a description of their main features, the two models are evaluated in terms of their outputs: what type of design-relevant information can be expected from a KLM analysis and from a BDM analysis? Whitefield stresses that both models are more evaluative than generative: they allow an assessment of an existing system, they do not permit generation of a new system. This limitation is not specific to KLM and BDM: evaluation is the main focus of most existing models. However, KLM and BDM do not provide the same type of system evaluation. KLM outputs a measure of system performance, in terms of time to complete the task. BDM outputs a model of the user, in terms of a prediction of the knowledge sources necessary to complete the task. KLM is oriented towards a metric of performance, while BDM is oriented towards a diagnosis on the activity. In this respect, BDM seems more suited to design, since it indicates possible points to be improved. It is, however, probably more difficult to use than KLM.

A possible solution to the failure of models to generate design ideas is addressed in the two chapters by Lansdale and by Falzon. These authors consider the hypothesis that observations of human behaviour in analogous, non-computerized tasks can provide useful information for designing the computer version of the software. In other words, the authors attempt to extract behavioural models from the observation of subjects fulfilling non-computerized tasks. Lansdale's domain of interest is the functioning of human memory (coding and retrieval processes) and its application to filing and retrieving files. Falzon analyses the lessons that can be learnt from human–human dialogues for the design of human–computer dialogue strategies. Some caution is necessary when one wants to follow this approach. Starting from Malone's observations on office organization (Malone, 1983), Lansdale stresses that the transfer between the two situations is not a straightforward process:

- The observed behaviour may not be optimal. For instance, Malone observes that subjects, failing to classify some items, create unstructured piles. This does not mean that computers should allow the creation of such piles. Transferring a deficient behaviour is non-productive.
- The observed strategies are the result of an interaction between man and work environment. What is observed is the behaviour of a system (in Whitefield's sense). For instance, the physical formats of the documents make it possible to stack them in piles; consequently, the subjects may retrieve documents by scanning the piles (using physical characteristics of the documents), by remembering their location, by focusing the search on some levels of the piles (since piles are implicitly ordered in a temporal order). Subjects compensate their lack of organization in classification by using appropriate retrieval strategies.

Thus the transfer cannot be straightforward. The 'pile' concept cannot be transferred without an understanding of the strategies used by the subjects and the document descriptors they imply. It is the psychology underlying behaviour that has to be elucidated.

In the same way, Falzon reports a number of studies that have attempted to measure the applicability of the human–human communication model to human–computer interaction. Results indicate that human verbal behaviour varies according to the model of the interlocutor. When the interlocutor is a machine, subjects' behaviours vary according to the representation they have of the machine. This representation may be induced by the machine behaviour. The problem is then to choose a machine behaviour that provokes a user behaviour compatible with the machine linguistic abilities and domain competences. Again, users' behaviour cannot be analysed independently of environment.

Lansdale and Falzon use different domains to exemplify the approach they advocate, i.e. an approach that allows one to reach a level of description which is both psychologically valid and useful for the design of interactive systems. The illustration proposed by Lansdale concerns the choice of file attributes to facilitate information retrieval. Falzon explores several aspects of human–human dialogues: the language and the mode of interaction, the effects of communication competence on dialogue strategies, and different aspects of cognitive economy in dialogue.

REFERENCES

Card, S. K., Moran, T. P and Newell, A. (1983). *The Psychology of Human–Computer Interaction*. Hillsdale, NJ, USA: Erlbaum.

Malone, T. W. (1983). How do people organize their desks? Implications for the design of office
    information systems. *ACM Transactions on Office Information Systems*, 1(1), 99–112.
Polson, P. G. and Kieras, D. (1985). An approach to the formal analysis of user complexity.
    International Journal of Man–Machine Studies, 22, 365–94.
Whitefield, A. (1989). Constructing appropriate models of computer users: the case of
    engineering designers. In: *Cognitive Ergonomics and Human–Computer Interaction*. J. B.
    Long and A. Whitefield (eds). Cambridge, UK: Cambridge University Press.

HUMAN–COMPUTER INTERACTION MODELS AND THEIR ROLES IN THE DESIGN OF INTERACTIVE SYSTEMS

Andy WHITEFIELD

Ergonomics Unit, University College London,
26 Bedford Way, London WC1H OAP, UK

1 INTRODUCTION

Models abound in the human–computer interaction (HCI) literature. They are common as ways of describing users, or computers, or tasks, or interactions, indeed, of describing any aspect of a human–computer system. Presentation of a model in the literature is often accompanied by the suggestion that it could be useful in system design. In too many cases this is simply said in genuflection to the desire for applicable research—the major application of HCI research knowledge being to the design of improved systems. Many authors do not suggest *how* the model might be useful in design, and very few illustrate, much less demonstrate, that usefulness.

This chapter examines critically the potential use of HCI models in the design of interactive computer systems. Two necessary initial steps are to consider the classes of HCI models that exist and the nature of system design. The classes of model that might be useful in system design are then outlined and the roles of two particular models described. Finally, these roles are considered from a number of perspectives.

2 THE RANGE OF HCI MODELS

While HCI models are clearly heterogeneous, what their salient characteristics and relationships are is less clear. Long (1987) has illustrated the confusion surrounding the nature of many HCI models by comparing how different authors refer to the same models. Here, a classification of HCI models is proposed in an attempt to clarify the different types of model and their relationships. Further details can be found in Whitefield (1987).

COGNITIVE ERGONOMICS:
UNDERSTANDING, LEARNING AND DESIGNING
HUMAN–COMPUTER INTERACTION

2.1   A Classification of HCI Models

The classification is intended to cover those models in the HCI literature that are expressed at the level of individual users. These models are typically called user models or user's conceptual models—general terms which themselves contribute to the confusion on the topic. Perhaps the most important problem to be faced in developing a classification is its descriptive basis; that is, how to determine the appropriate discriminations between the elements? In their discussion of classification in the area of human performance, Fleishman and Quaintance (1984) suggest that one way to proceed is to define the elements to be classified. They say a good definition 'will permit the derivation of terms that reliably describe tasks and distinguish among them. These derived terms provide the conceptual basis for classification' (p. 49).

Although some definitions will be better than others, their differing purposes mean that there will not be a 'best' definition; a definition might be well-suited to meeting one requirement (e.g. classification) but not another (e.g. model construction). Fleishman and Quaintance suggest that one should 'adopt or develop a definition that will serve as an adequate vehicle for classification' (p. 49). On this basis, the definition to be used here is the following: a model is a characterization of an X held by a Y. This definition allows the derivation of two terms as the basis for a classification: X—who or what is being modelled? and Y—who or what is doing the modelling? Clearly this definition is weak. It fails to address several potentially important aspects of models, such as their purpose and the nature of the relationship between the characteristics of X and Y (see, for example, Long, 1987; Streitz, 1988). But firstly, the definition is adequate and appropriate for the purpose of discussing the roles of models in design, as hopefully will be clear below. And secondly, if the definition should prove inadequate, it is certainly extensible—one could easily add to it in a way that would allow the derivation of more terms.

So what might be values for the X and Y terms? In his discussion of what he calls the user's conceptual model, Young (1983) notes that there are four agents who might be involved in developing such a model; with some substitution of terms, these are the *user*, the *designer*, the *researcher* and the *program*. These four agents will serve here to detail the Y term in the above definition. The terms user, designer, researcher and program will often be referred to by the letters U, D, R and P respectively. The term program will be used to refer to all the software involved in a particular interaction. 'Computer' would be an alternative term here, but program is preferred because it focuses on software and minimizes hardware considerations. In computer contexts the term system is usually used in the sense of program here, i.e. to refer to a collection of software components; but system will be used here in its ergonomic sense, to refer to the combination of user and program. A system (S) is thus an artefact arising from the interaction between the user and the program for a given task; it is not an agent in that interaction and cannot itself develop models.

Concerning the X term in the definition, who or what might these four agents model? In principle, the same four could all be the subject of the others' models; in addition, each could model the system. This produces a set of five possible entities to be modelled (user, designer, researcher, program and system). These distinctions can therefore be used to classify the models found in the the HCI literature into twenty classes, as shown in Table 1. In our current state of knowledge, some classes (particularly models of researchers and designers) might appear to be of limited relevance, and examples of them may not yet exist. But all classes are legitimate and may be required in the future. Models in the majority of classes can be identified now. Some will be used as illustrations.

Table 1. Classification scheme for human–computer interaction models

Model Of / Modelled By	SYSTEM	PROGRAM	USER	RESEARCHER	DESIGNER
PROGRAM					
USER					
RESEARCHER					
DESIGNER					

Various aspects of this classification need discussion, starting with two general points. Firstly, models of tasks are not treated as a separate class but as instances of user, program or system models. Thus tasks are not considered to have an existence independent of a system or system component. Secondly, the Y term in the definition (who is doing the modelling) encompasses both the creator and the utilizer of a model. In most cases these roles coincide. Occasionally they do not; for example, most designers' models are created by researchers for use by designers.

For the purposes of this chapter, it is convenient to consider the classes in two groups: (a) programs' and users' models; and (b) researchers' and designers' models. In each case the discussion will focus on those classes for which example models do currently exist. The classes may be referred to by the letters denoting their two agents, e.g. R(P) denotes a researcher's model of a program.

2.2 Programs' and Users' Models

General introductions to notions of a program's model of a user are provided by Rich (1983) and Murray (1987). Various forms of such a model are possible, ranging from direct manipulation interfaces having additional keyboard menu selection for experts, to sophisticated computer-based learning programs which adapt the dialogue according to the state of the session and its progress. Such models are potentially very important in improving interactions, but because they are intended as part of the program, rather than for use in system design, they are of limited interest here and will not be discussed further.

What constitutes a user's model of the system or program is probably the single most problematic distinction in the literature. As used here, a U(P) is *not* the mental representation of the program that is literally in the user's head. Such a model could never be represented externally or by others, except as a model of that model. Under such an interpretation, therefore, one could never show an example U(P) and the category would always be empty; further, any classification would inevitably be required to include higher-order models (e.g. A's model of B's model of C—see Streitz, 1988).

The alternative interpretation used here is that a U(P) is a model which attempts to be psychologically accurate, at a detailed level, about the model of the program held by the user. That is, it tries to represent the user's actual mental structures and processes relating to the program. The extent to which a model is psychologically accurate is of course more a continuum than a dichotomy. Perhaps the easiest (but hardly the best!) way to identify U(P)s is by their irregularity. Norman (1983, p. 14) warns that they are likely to be 'messy, sloppy, incomplete, and indistinct structures' (well illustrated by Hammond et al., 1983) and discovering their details is a difficult task requiring psychological experimentation and observation of users. Most models (e.g. those discussed by Young, 1983) are not meant to be psychologically accurate in this way, although their contents may be closely related to the contents of U(P)s. Rather, they attempt to characterize and perhaps predict the user's behaviour, usually stressing its regularities.

Given this interpretation of U(P), an example is the information structures model of Morton et al. (1979). Others are part of the task/action mapping of Young (1981), the external/internal task mapping (ETIT) of Moran (1983), and the goal structures model of Morton et al. (1979). These three are examples of models that, although presented as single models, really comprise two separate models with a defined relationship. The two constituent models are usually of the program and of the user. They differ from models of the system in that the user and program elements of the system are described separately (although often in the same representation).

2.3 Researchers' and Designers' Models

Researchers' and designers' models will be considered together here (and throughout the chapter) because they are both currently almost exclusively developed by researchers, as stated above. Not surprisingly, researchers' models have received the most attention in the literature. Relatively little effort has been put into investigating models either held by designers or constructed specifically for them. Whether researchers' and designers' models could or should be the same is a question for debate.

The most detailed models of the system (i.e. the combination of user and program) are the researchers' models of text-editing known as GOMS, and their associated designers' models, the Keystroke-Level and Unit Task-Level models (Card et al., 1983). A given GOMS model is of a particular system, so its content is dependent upon both the user and the program. A model based on a line-editor will therefore differ from one based on a screen editor even for the same task. The Keystroke-Level and Unit Task-Level models predict system performance in terms of time on task for an expert user. The idea is that the designer can compare and choose between different possible systems using this metric.

Another R(S) is presented by Reisner (1984). She describes an 'action grammar', in a BNF-style notation, which has as terminal symbols both cognitive actions (e.g. retrieve from working memory) and physical actions (e.g. keystrokes). The former clearly describe the user, while the latter really describe not the user but the program in terms of strings of legal command inputs. This form of the grammar contrasts with Reisner's earlier (1981) version, where all the actions were physical ones. Since it describes the command language, that earlier version is classified here as a model of the program (see below).

If one does not have a model of the system, then, in considering how users would interact with a program, one must have some kind of R(P) or D(P) model. Designers already operate with several models of the program (e.g. modular trees, flowcharts, functional descriptions—see Shooman, 1983). The relevant question here is how appropriate these models are for addressing human factors issues. Embley and Nagy (1981) experimented with both grammars and state transition diagrams as D(P) models. While both were useful, they were limited in the human factors issues they could address.

Examples of program models in the HCI literature include the second parts of the goal structures, ETIT, and task/action mapping models discussed above (describing the program), the generalized transition network of Kieras and Polson (1985), the surrogate model of Young (1981), and Reisner's (1981) version of a grammar.

The final types of model to be considered are researchers' and designers' models of the user. Since they can be concerned with modelling any aspect of the user's behaviour relevant to the task, they are likely to vary a great deal. This variety includes Whitefield's (1985, 1989) blackboard model of the engineering designer's activity (describing the classes of knowledge recruited during this particular task), and Card et al.'s (1983) Model Human Processor (a context-independent collection of perceptual, motor and cognitive processors, memory stores, operating times and principles of operation). These models address different aspects of the user and serve different purposes, as should be clear below.

There is evidence that designers can and do operate with some general models of the program's users (e.g. Hammond et al., 1983). It is the inadequacy of these models that is often given as a principal reason for human factors difficulties (e.g. Dagwell and Weber, 1983). Current D(U)s are often implicit, ill-formed and inaccurate. HCI researchers have been a little slow to propose specifically D(U) models, both because their relationships to programs can be unclear, and because there are important questions of the form they should take to be useful and usable.

One point worth noting about models in general is that there appears to have been a move away from models of isolated system components (i.e. of users and of programs) and towards models which include both components (i.e. models of systems or related models of users and programs). Examples of this trend are Reisner (1981, 1984), who has developed her original BNF grammar model of the program into a model of the system, and Whitefield, who has developed an additional model of the program to be related to the original blackboard model of the user (Whitefield, 1986; Long et al., 1988). This will be discussed in more detail later. This trend is clearly a good thing insofar as it comes closer to the requirements for the design of interactive systems, as should be clear later in Sections 3 and 4.

2.4 Discussion of the Classification

As a whole, the classification shows promise in discriminating and relating HCI models. Some simple but useful distinctions have been made to clarify different types of model. The two dimensions to the classification admit a more principled view of the differences between models than one or more single discriminations, such as those proposed by Norman (1983). In addition, both the distinction between the program and the system, and the identification of some models as composed of two constituent models, allow a better appreciation of the scope of any model.

Similar ideas have been put forward independently by Streitz (1988). The classification here has the advantages of a more explicit derivation, of the system/program distinction, and of remaining somewhat simpler; but Streitz's discussion of second (or nth) order models will have benefits for some purposes.

There are of course a number of problems with the classification. One is that not all the relevant agents are identified. As mentioned earlier, the classification fails to distinguish a model's utilizer from its creator (Long, 1987). Similarly, both computer scientists and human factors practitioners can be researchers. A second problem is that most classes are themselves heterogeneous, containing some models with few similarities. Undoubtedly, further classification of the models would be productive, but it may need to be done on different bases for different classes or group of classes.

Further classification would also involve extending the model definition. As pointed out in Section 2.1, this does not capture some relevant aspects of models, although it is extensible. This might involve including higher-order models, although the need for this hinges around a further problem—the distinction between U(P) and other models. The notion of psychological accuracy must be better specified to be reliable.

It is intended that this classification, by identifying different classes of model, will serve as a basis for considering how models might be used by designers in developing interactive systems.

3 THE DESIGN OF HUMAN–COMPUTER SYSTEMS

Before considering how the types of model identified might be used in the design of interactive systems, it is necessary to consider the general nature of that design activity. The study of design is a wide-ranging cross-disciplinary field with its own literature and research interests, which it would be impossible to summarize here. Instead, the design studies literature will be drawn upon to make a few points about system design which are particularly relevant for the purposes of this chapter. The general aim is to identify some important features of the design of interactive systems and thus to provide a context for the discussion of the use of models in design that follows.

The first point to be made is that design is not a logical and deterministic activity. This has been argued theoretically by Cathain (1982) and the earliest empirical investigations of design showed that, in practice, the solution structure does not follow logically from an analysis of the problem (see for example Cross, 1984). This is because design problems are complex, inevitably under-specified, and admit a range of acceptable solutions rather than a single optimum solution. As opposed to the deterministic view, design is more of a dialectic between the generation of possible solutions and the discovery of the constraints operating on the problem and solution spaces.

The second point concerns useful ways of describing the design activity. Most descriptions of design are of two kinds: either they involve some stage-based description of the life cycle, or they focus on the processes involved throughout. Life cycle descriptions tend to be idealized. They usually bear strong similarities to each other, varying mostly about when they think design starts and finishes. For current purposes, and based on Walsh et al. (1988), design will be considered as a two-stage activity. The first stage takes a statement of the customer's needs and requirements and produces a functional specification of the system; this involves defining the abstract software components in a suitable notation. The second stage instantiates this functional specification, i.e. produces a working system implemented in some form; this stage involves designing and coding the components.

There are a number of process descriptions of design, many of which have a great deal in common. Most of the early ones were domain-independent and prescriptive, in that they proposed how design should be carried out. They tended to distinguish the (supposedly sequential) processes of *analysis* (of the problem), *synthesis* (of a solution) and *evaluation* (of the solution adequacy). Recent process models are more empirically based and descriptive of actual design behaviour. Whitefield (1989) has used the blackboard architecture from artificial intelligence to propose a model involving the activities of *generation, evaluation* and *control* (of generation and evaluation). The analysis process of the earlier models is here divided between the three processes. While these process descriptions have not focused on interactive system design, they may be applied to it and will serve here to characterize it.

What is the relationship between the life cycle and process views? Although early writings attempted to link particular processes strongly with particular stages, the developing dialectical view of design outlined above suggests that all processes are used at all stages. This is openly acknowledged by Long and Whitefield (1986) who, in their consideration of evaluating interactive systems, state that evaluation can take place at all stages of system design. This contrasts with the (mostly implicit) common view that evaluation can take place only when some implemented form of the system exists, even if it is a simulation or a prototype.

The third point to be made about design concerns what might be called alternative design approaches. Often discussion in the HCI literature about human factors inputs to design is predicated upon assumptions that design proceeds, or ought to proceed, in a particular way. Broadly speaking, the two major approaches discussed can be characterized as rapid iterative prototyping (with evaluative feedback on each prototype being incorporated into the next version) and structured or formal methods (where formal is interpreted broadly to include any method able to ensure the development of systems which meet some criterion of effective performance—see Long and Dowell, 1988). Although these approaches are not in principle mutually exclusive, current practice usually either assumes they are or operates as if they were.

The purpose here is not to discuss the relative merits of these (or other) approaches. Local circumstances (e.g. type of system, its size, whether original or variant design, available resources, access to users, organizational practice, etc.) will tend to determine, or at least influence, the approach adopted. The point is that each approach entails a different scope for the use of models. In particular, since a working version is available very early, the prototyping approach will tend to be concerned with users' models of the program or system, based on empirical testing of the current version. In contrast, the structured approach will be more concerned with designers' models of the system, user and program; these can be examined or manipulated to make predictions about the system in advance of its implementation. This is what Reisner (1983) calls the use of models as analytic tools. Moreover, the structured approach allows for this recruitment of the different models in an orderly fashion. The focus in sections 4 and 5 will be on the roles of models within the structured approach.

The final point to be made here about design concerns what is being designed, and what it is meant to do. The use of the term 'system' in two ways, as covered in section 2, has probably contributed to some lack of clarity as to what system design means. Often it is taken as meaning the design of only the computer component of a human–computer system; but it should be taken as meaning the design of both human and computer components. This is clear if one considers HCI as a discipline, as is done by, for example, Dowell and Long (1988). They attempt to clarify HCI concepts and their relationships. They describe systems in HCI as human–computer systems performing tasks which produce changes in the objects constituting the world of work. It is therefore necessary that design concerns the complete system, and consequently any aids to design, such as models, must have an explicit relationship with the complete system.

An additional aspect of the system focus concerns what the system does. According to Dowell and Long (1988), systems perform tasks. This performance is expressed in terms of the quality of the task outcome (e.g. the quality of a letter produced using a word processor) and the cost to the system (e.g. time taken). Performance is determined by behaviour, i.e. *how* the system accomplishes its task. The system's behaviour would thus concern the representations, structures, and processes recruited, objects manipulated, actions undertaken, and so on. This distinction between a system's performance and its behaviour relates strongly to how models might be useful in design, and is discussed below.

4 THE ROLES OF MODELS IN DESIGN

The description of the various model types and the consideration of design now allows a discussion of how various models might contribute to design. This section will firstly discuss this in terms of model classes, and will then describe the applications of two particular models.

The first point to be made concerns the role in design of users' models of programs or systems, i.e. of U(P) or U(S) models. These can really only be developed when users have a program to use, and they will therefore be most effective in the manner of 'error ergonomics' (Singleton, 1971). That is, the problems and difficulties highlighted in the models (probably by their consistency and complexity) can be fed back to the designer to improve the next version of the program. Such models would therefore be appropriate for the iterative prototyping approach to design outlined earlier. The structured approach allows more scope for the use of models, and will be the focus of the discussion from here on. Users' models will not be discussed further.

As outlined in section 3, it should be axiomatic that interactive system design necessitates consideration of both system components—the user and the program. This means that the models most useful to designers will be those which include both components—either models of the system (e.g. GOMS) or related models of the user and program (e.g. ETIT).

Since individual models of the user or of the program only model half the system, their usefulness to the designer is less. To use a model of half a system to make performance predictions for the whole system requires assumptions about its relationship to the other system component. At best, these assumptions will be clearly specified and explicitly derived; at worst, they will be unspecified and implicit. An example where the assumptions are relatively explicit is Reisner's (1981) earlier model based on BNF grammar. This was classed above as a researcher's model of a program, and Reisner has tried to demonstrate how it might be used to predict some performance difficulties. But two sorts of assumption are involved that constitute a match with an implicit view of the user. The first is that there is a similarity between the user's mental representation and the model. The second is about the desirable psychological features of the model; for example, memory limitations are given as the basis of the assumptions that few rules and a large number of terminal symbols are desirable. A good deal of psychological knowledge is required to make accurate assumptions of this kind.

An example of a model of a program where the assumptions are much less explicit can be the use of metaphors or analogies. Designing a program that represents an office desktop involves a model of the program that incorporates assumptions about how users will learn and operate it.

Similar considerations apply to isolated models of the user in design— assumptions must be made about the interaction with the program half of the system. To point out the need for these assumptions is not to claim that these single model approaches cannot be useful. Reisner's grammar probably would be useful in the absence of any more system-oriented predictions about errors. Hammond et al. (1983) report that designers do use their own ill-specified models of users, which they relate in their own way to the program, and these models can be the bases of design decisions. The principal point is to recognise that the assumptions are involved, and to attempt to be as clear as possible about them. Of course, the better specified they become, the more the relevant model will resemble a model of the system.

The preferred approach to using models in design, therefore, involves models which explicitly include both system components. Two examples of models suggested for designers' use will now be described, each of which models the complete system in some form. The comparison of these will be the main basis of the discussion of model roles in section 5. The two examples have been selected because their authors have written about their possible use by designers, and because the contrasts between them are both numerous and clear; they should therefore provide the most appropriate and informative contrast. They are the Keystroke-Level Model (Card et al., 1983) and the Blackboard Design Model (Whitefield, 1989). These will be referred to as the KLM and BDM respectively.

The KLM is a simplified version of the GOMS model of text editing. It describes system performance in terms of time on task for an expert user of a given program on a given task using a given method. It contains a set of operators (at the level of keystrokes, pointing, mental operations, and so on) organized as a method (i.e. a sequence of operators for accomplishing a task). Given the command language of the program, the designer can construct methods for any task. A set of rules assigns operators to each step of a specified task. Then standard times assigned to the operators enable the calculation of predicted performance times for each task. The designer can compare this with some benchmark or criterion, or use it to select between alternative possible systems.

For example, a designer might have to decide on the syntax for a REPLACE command on a word processor. One option is to point at the word to be replaced, then to type the command and the replacement word; the second option is to type the command, then to point at the word, then to type the replacement word. A KLM model of these options suggests that the former would be performed faster, and therefore be preferred, largely because less time is spent moving between the input devices.

Card et al. are very clear about the deliberately quantitative orientation of the KLM. They suggest that just as the engineering designer uses calculational formulae, the interactive system designer should do the same. 'The ability to do calculations is the heart of useful engineering-oriented applied science. Without it, one is crippled' (1983, p. 10).

The role of the KLM is thus as an evaluative metric of system performance. It has a clear analogy with the use of, for example, daylight factor indices in architecture and stress analyses in civil engineering. It has a routine application method (indeed it clearly has the potential for automated application) producing a single result. The designer is required to have little or no understanding of user behaviour to apply it.

The BDM has been developed in a different application domain—that of computer-aided design (CAD) systems. Its use involves separate user and program models. The model of the user (the BDM itself) gives a decomposition into approximately twenty classes of the knowledge applied by a mechanical engineering designer during a design task. These knowledge sources can be grouped into four types (generative and evaluative domain knowledge, and generative and evaluative operating knowledge). The model provides an architecture for the structuring and application of this knowledge, in a form derived from the blackboard models of artificial intelligence (Hayes-Roth, 1983). Thus the model describes the knowledge the designer brings to bear upon the task plus the way in which this is done. Both these static and dynamic aspects will vary for different tasks and for different designers.

In application, the BDM is mapped with a model of the CAD program (Whitefield, 1986; Long et al., 1988). This program model was specifically developed for this purpose and focuses on the program's functions. These are divided according to whether or not they concern the object specifications (i.e. the product of the design) and the types of operations they perform (development, interrogation, display, management). Further divisions can be made for particular programs.

The intended application of the BDM is that the system designer should develop models of the user and program for the system being developed. The designer would then map the two models by reasoning about how the various program functions interact with and support the user's knowledge recruitment. This mapping is done both structurally (i.e. identifying the static entities in each model) and procedurally (i.e. identifying constraints placed in use on knowledge source or function ordering). Relevant criteria about what constitutes an appropriate and desirable use of knowledge would enable decisions to be made between possible systems.

For example, at the level of groups of functions and knowledge sources, the amount of support offered by the program's interrogation and display functions to the user's evaluative domain knowledge might be inadequate; at a more detailed level, the primitives within the program's development functions might not match the primitives recruited by the user's generative domain knowledge. In each case, criteria about the use of domain knowledge might enable decisions between possible system organizations to be made.

The role the BDM plays within system design is therefore different from that of the KLM. The BDM is intended to allow the assessment of user/program compatibility, by predicting, for a given system configuration and task, the behaviour of the system in terms of the recruitment of user knowledge and program functions. Note that this role is still primarily evaluative (albeit of a different kind) rather than generative. It is not part of the model's application to propose particular system changes, although clearly the assessment will be fertile ground for identifying possible beneficial changes.

The use of this type of model demands considerable effort and skill from the designers. They would need to learn about the structure and operation of the model under different circumstances. It could not be applied in a rote manner and, certainly in its current form, a good deal of scope for interpretation is available.

Other HCI models for use by designers tend to fall between these two contrasting alternatives. For example: the calculation of scrolling times used to illustrate Reisner's (1984) system model, and Moran's (1983) proposed use of metrics to quantify the difficulty of mapping user and program models in ETIT, both resemble the performance quantification of the KLM; but identifying mismatches in the goal structures of the user and program (Morton et al., 1979), and developing core mappings between the task and action domains (Young, 1981) have more in common with the qualitative behavioural descriptions of the BDM.

5 DISCUSSION OF THE ROLES OF MODELS IN DESIGN

Clearly the proposed design roles of the KLM and BDM differ in a variety of ways, a number of which will be discussed here. These different perspectives will serve to clarify the nature of these design roles.

The first perspective (which in fact tends to emphasize the similarity between the models rather than their differences) is to ask how the roles relate to the processes involved in design. As stated in section 3, the processes most commonly identified are generation (or synthesis) of solution elements and evaluation of solution adequacy. In these terms, both the KLM and BDM constitute evaluations. The KLM supports an evaluation of system performance in terms of time on task, while the BDM supports an evaluation of system behaviour in terms of the recruitment of user knowledge and program functions. In both cases, the evaluations could be made against benchmarks (e.g. performance time must be less than T; knowledge recruitment must be at least K) or they could be compared for alternative system versions. But neither model operates directly as a generative process to propose elements in the system design.

This therefore raises the question of whether *any* models could fulfil a generative rather than an evaluative function. There is at least a strong case to be made that models can only be evaluative. This is based on the non-deterministic nature of design discussed earlier, which suggests that generating the detailed content of a system from a less detailed model of it is logically impossible. Indeed, the KLM's authors make a similar point. They say that performance models fulfil an evaluative function, but cannot fulfil a generative one: 'For any interesting real-world domain of design, there cannot be any global synthesis function that maps requirements into a structure' (Card et al., 1983, p. 406). Against this case, there are some models for which a more generative role has been suggested. For example, Johnson (1985) has proposed that with TAKD (Task Analysis for Knowledge Descriptions), the entities arising from a task analysis could be directly utilized as the objects in an object-oriented system.

There is currently insufficient experience in the use of HCI models in design to decide whether they can be generative or only evaluative. One possibly important factor is the stage of design at which they are used. Both the KLM and BDM would probably be applied during the second stage of implementing the functional specification, whereas TAKD might relate more to the first stage of moving from the requirement statement to the functional specification. Alternatively, it may be that whether one sees a generative or solely evaluative role for models depends upon one's exact interpretation of the two processes, which may be rather blurred at the early stages of design. At the moment, however, the view that models support evaluative processes in design is highly consistent with the state of the art, and also ties in with the view that evaluation can be done at any time during design (Long and Whitefield, 1986).

The second perspective on the model roles follows on from this. Although both example models play evaluative roles, it is clear that the BDM is likely to be more productive for considering possible additional or alternative elements to the design solution than is the KLM. For example, the BDM can be used to identify that certain of the user's knowledge sources are unsupported by program functions, or that the program's primitives are at too low a level relative to the user's. Evaluations of this kind clearly indicate areas where the design must be improved, although they do not directly suggest particular solution elements. With the KLM, in contrast, the output is a numerical prediction of performance time, which does not itself highlight areas for improvement. Certainly inspection of the model might identify problematic areas, but this is not its basic mode of application. In the terms of Long and Whitefield's discussion of evaluation, the BDM is *diagnostic* about the system while the KLM undertakes *measurement*. Diagnosis is more useful information for the developers in improving the system design.

This effect could be due to a number of differences between the models, such as their level of detail, or their task scope. But the underlying reason seems to be that the KLM addresses system performance (how long it takes to carry out a task) whereas the BDM addresses system behaviour (how the task is carried out). It is the focus on behaviour rather than performance that allows the BDM to be more productive in considering alternative solution elements. It is not the case that a model will exclusively address performance or behaviour, but that its primary intended output will be in terms of one or the other. As far as the designers are concerned, a requirement for diagnostic evaluation would be an important factor in selecting models for application.

Similar distinctions have been made before, suggesting that this is a critical aspect of HCI models. For example, Williges (1987) divides models into conceptual (concerned with cognitive processes, structures and strategies) and quantitative types; Whitefield (1986) used the terms compatibility and performance; and there is also a strong overlap with the distinction between qualitative and quantitative models. The performance/behaviour distinction has the advantage of its orientation towards systems, rather than just users, and its grounding in systems theory.

A third perspective on the models' design roles overlaps with this performance/behaviour difference; this is the clear contrast between the KLM and BDM in their ease of use. The KLM appears easy to use; it has a straightforward application and an apparently meaningful quantitative result (although this does require interpretation in terms of the complete system design). The BDM seems much harder to use; it has a flexible application demanding considerable understanding on the designer's part, and produces qualitative analyses of system behaviour that require further interpretation. These differences in effort will probably tend to correlate with performance/behaviour models, but will not always do so. ˙

The alternatives correspond respectively to what Harker and Eason (1984) call the 'information acquisition' and 'knowledge transfer' approaches to the use of ergonomic knowledge by designers. Information acquisition approaches (e.g. guidelines and checklists) try to provide designers with specific information that requires little understanding of user behaviour to be applied. Knowledge transfer approaches, on the other hand, involve designers learning human factors skills. These may be either methodologies like task analysis, or technical expertise in particular aspects of user behaviour.

Each approach has its own disadvantages. The knowledge transfer approach is time-consuming and risks unskilled use resulting in misinterpretation both in its application and in assessing its outcomes. The information acquisition approach risks being misleading in its apparent quantification and being too simplistic and restricted for the complexity of human–computer interactions (rather like work-study approaches to human–machine interaction). This approach tends to treat designers as if all they can use is a tool with a straightforward and prescribed application, rather than as being able to make intelligent use of the available information. Local circumstances will tend to determine the use of either approach, but they ought to be viewed as complementary rather than as alternatives.

Indeed, this last point is an important general point and an appropriate one on which to end this discussion of model roles. Interactive systems are complex entities, and it would be unrealistic to expect only a single type of model to be useful in their design. As Williges (1987, p. 501) puts it: 'A rigid adherence to one approach or one specific model is counter-productive to understanding the true complexity of the user interface'.

The three general perspectives outlined here, in terms of the design processes the models support, their focus on system performance or behaviour, and their ease of use, have been the most fruitful ways in which to clarify the roles of HCI models in design. A number of other differences between the KLM and BDM model roles can be identified (e.g their level of detail); but they are more difficult to use as general comments on the roles because their sources are unclear (e.g. just between the particular models, or the model classes, or the application domains, or whatever).

6 CONCLUDING COMMENTS

This chapter has sought to examine the potential use of HCI models in the design of interactive systems. Classifying the models and highlighting important features of system design allowed the identification of classes of model that could be useful in design. These are principally users' models of programs or systems, for an iterative prototyping approach to design, and designers' models of systems (or related models of users and programs) for a more structured approach to design. Within the latter approach, a comparison of Card et al.'s (1983) Keystroke-Level Model and Whitefield's (1989) Blackboard Design Model, has suggested that there are similarities and differences in the roles models might play. Although they support design evaluation processes, other aspects, and particularly their orientation towards system performance or behaviour, will be an important determinant of how designers might utilize them.

ACKNOWLEDGEMENTS

I would like to thank Norbert Streitz and my colleagues at the Ergonomics Unit who commented on an earlier version of this chapter.

REFERENCES

Card, S. K., Moran, T. P. and Newell, A. (1983). *The Psychology of Human–Computer Interaction*. Hillsdale, NJ, USA: Erlbaum.

Cathain, C. S. (1982). Why is design logically impossible? *Design Studies*, 3(3), 123–5.

Cross, N. (ed.) (1984). *Developments in Design Methodology*. Chichester: Wiley.

Dagwell, R. and Weber, R. (1983). System designers' user models: a comparative study and methodological critique. *Communications of the ACM*, 26(11), 987–997.

Dowell, J. and Long, J. (1988). Towards a paradigm for human–computer interaction engineering. In: *Contemporary Ergonomics 1988*. E. D. Megaw (ed.). London: Taylor & Francis.

Embley, D. W. and Nagy, G. (1981). Empirical and formal methods for the study of computer editors. In: *Computing Skills and the User Interface*. M. J. Coombs and J. L. Alty (eds). London: Academic Press.

Fleishman, E. A. and Quaintance, M. K. (1984). *Taxonomies of Human Performance: the Description of Human Tasks*. Orlando, FL, USA: Academic Press.

Hammond, N., Jorgensen, A., MacLean, A., Barnard, P. and Long, J. (1983). Design practice and interface usability: evidence from interviews with designers. In: *Proceedings of CHI '83*, pp. 40–44.

Hammond, N., Morton, J., MacLean, A. and Barnard, P. (1983). Fragments and signposts: users' models of the system. *Proceedings of the Tenth International Symposium on Human Factors in Telecommunications, Helsinki*.

Harker, S. D. P. and Eason, K. D. (1984). Representing the user in the design process. *Design Studies*, 5(2), 79–85.

Hayes-Roth, B. (1983). The blackboard architecture: a general framework for problem solving? *HPP Report No. HPP-83-30*, Stanford University, Computer Science Dept.

Johnson, P. (1985). Towards a task model of messaging: an example of the application of TAKD to user interface design. In: *People and Computers: Designing the Interface. Proceedings of HCI '85*. P. Johnson and S. Cook (eds). Cambridge: Cambridge University Press.

Kieras, D. and Polson, P. G. (1985). An approach to the formal analysis of user complexity. *International Journal of Man–Machine Studies*, 22, 365–94.

Long, J. B. (1987). A framework for user models. In: *Proceedings of the Ergonomics Society Annual Conference, 1987*. London: Taylor & Francis.

Long, J. B. and Dowell, J. (1988). Formal methods: the broad and the narrow view. *IEE Colloquium on Formal Methods and Human–Computer Interaction*, February 1988, London.

Long, J. B. and Whitefield, A. D. (1986). Evaluating interactive systems. *HCI '86* September 1986, University of York. (Tutorial given).

Long, J. B., Whitefield, A. D. and Warren, C. P. (1988). Final report to SERC on grant reference GR/D/72969 (Alvey reference MMI/142).

Moran, T. P. (1983). Getting into a system: external–internal task mapping analysis. In: *Proceedings of CHI '83*, pp. 45–9.

Morton, J., Barnard, P., Hammond, N. and Long, J. (1979). Interacting with the computer: a framework. In: *Teleinformatics '79*. E. J. Boutmy and A. Danthine (eds). Amsterdam: North-Holland, pp. 201–8.

Murray, D. M. (1987). Embedded user models. In: *Human–Computer Interaction—Interact '87*. H-J. Bullinger and B. Shackel (eds). Amsterdam: North-Holland.

Norman, D. A. (1983). Some observations on mental models. In: *Mental Models*. D. Gentner and A. L. Stevens (eds). Hillsdale, NJ, USA: Erlbaum.

Reisner, P. (1981). Formal grammar and human factors design of an interactive graphics system. *IEEE Transactions on Software Engineering*, SE-7(2), 229–40.

Reisner, P. (1983). Analytic tools for human factors of software. In: *Enduser Systems and their Human Factors*. A. Blaser and M. Zoeppritz (eds). (Lecture Notes in Computer Science No. 150). Berlin: Springer-Verlag.

Reisner, P. (1984). Formal grammar as a tool for analyzing ease of use: some fundamental concepts. In: *Human Factors in Computer Systems*. J. C. Thomas and M. L. Schneider (eds). Norwood, NJ, USA: Ablex.

Rich, E. (1983). Users are individuals: individualizing user models. *International Journal of Man–Machine Studies*, 18, 199–214.

Shooman, M. L. (1983). *Software Engineering*. London: McGraw-Hill.

Singleton, W. T. (1971). Psychological aspects of man–machine systems. In: *Psychology At Work*. P. B. Warr (ed.). 2nd edn. Harmondsworth, UK: Penguin.

Streitz, N. A. (1988). Mental models and metaphors: implications for the design of adaptive user–system interfaces. In: *Learning Issues for Intelligent Tutoring Systems*. H. Mandl and A. Lesgold (eds). New York: Springer-Verlag.

Walsh, P., Lim, K. Y., Long, J. B. and Carver, M. K. (1988). Integrating human factors with system development. In: *Designing End-User Interfaces*. N. Heaton and M. Sinclair (eds). Oxford, UK: Pergamon Infotech.

Whitefield, A. D. (1985). A model of the engineering design process derived from Hearsay-II. In: *Human–Computer Interaction—Interact '84*. B. Shackel (ed.). Amsterdam: North-Holland.

Whitefield, A. D. (1986). Constructing and applying a model of the user for computer system development: the case of computer aided design. Unpublished PhD thesis, University of London.

Whitefield, A. D. (1987). Models in human computer interaction: a classification with special reference to their uses in design. In: *Human–Computer Interaction—Interact '87*. H-J. Bullinger and B. Shackel (eds). Amsterdam: North-Holland.

Whitefield, A. D. (1989). Constructing appropriate models of computer users: the case of engineering designers. In: *Cognitive Ergonomics and Human–Computer Interaction*. J. B. Long and A. D. Whitefield (eds). Cambridge, UK: Cambridge University Press.

Williges, R. C. (1987). The use of models in human–computer interface design. *Ergonomics*, 30, 491–502.

Young, R. M. (1981). The machine inside the machine: users' models of pocket calculators. *International Journal of Man–Machine Studies*, 15, 51–85.

Young, R. M. (1983). Surrogates and mappings: two kinds of conceptual models for interactive devices. In. *Mental Models*. D. Gentner and A. L. Stevens (eds). Hillsdale, NJ, USA: Erlbaum.

COGNITIVE COMPLEXITY OF HUMAN–COMPUTER INTERFACES:
An Application and Evaluation of Cognitive Complexity Theory for Research on Direct Manipulation-Style Interaction

J. E. ZIEGLER, P. H. VOSSEN and H. U. HOPPE

Fraunhofer-Institut für Arbeitswirtschaft und Organisation (IAO), Holzgartenstr.17, 7000 Stuttgart 1, FRG

1 INTRODUCTION

The development of workstations with multifunctional, integrated application software has led to attempts to improve the manner in which the user is able to access and use the different functions of such a system. The more the functionality of a system is broadened, the more the ease of learning is determined by the degree of consistency of use between the different functional domains.

Since the early 1980s a new generation of office workstations has become available (cf. Smith et al., 1983; Williams, 1983) which are trying to solve these usability problems by introducing a highly visualized, object-oriented interaction style between user and system. Shneiderman (1982) has coined the term 'direct manipulation' (DM) for this new mode of interaction. Although some authors have described characteristics of this interaction mode (Hutchins et al., 1986; Fähnrich and Ziegler, 1984) there are still many open usability issues related to this interaction style.

Usability is a multifacetted concept which has to be broken down into more elementary notions referring to observable and measurable phenomena at the human–computer interface. Also, these phenomena have to be related to psychological models of performance, learning, transfer of learning and development of cognitive skills and competence. Only in this manner can we establish predictive measures of usability which are needed to complement the costly empirical evaluation of prototypes and products. The development of such a methodological framework, particularly for the analysis of direct manipulation interfaces, forms part of the project HUFIT (Human Factors in Information Technology), which is pursued in the European ESPRIT programme.

This chapter deals with the establishment of predictive measures for learning and transfer of skills, the latter being a major indicator for the consistency of an interface. Cognitive complexity theory developed by Polson and Kieras (1985) offers a framework for the modelling of cognitive tasks and for deriving such predictions. In the following, we describe the background and features of this approach.

2 PRODUCTION SYSTEMS AS COGNITIVE MODELS

The term 'cognitive modelling' marks the efforts to construct computational models for mental processes (Johnson-Laird, 1983). Most of the current research in this field is done using production systems (PS) for the representation of procedural knowledge (cf. Anderson, 1983). A PS is a set of rules, each consisting of two components—a condition and an action part (Figure 1):

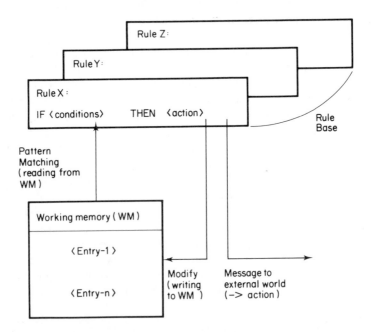

Figure 1.   The basic mechanism of a production system interpreter.

What makes PS rules much more flexible and powerful than IF–THEN-statements in classical programming languages is the interaction of the rulebase (set of PS rules) with a working memory (WM). The basic mechanism of this interaction is the recognize–act cycle. In the first step of a recognize–act cycle, the conditions of the currently used rules are matched against the WM entries in order to determine the possibly applicable rules. A conflict resolution mechanism decides, which of the selected rules will 'fire' (i.e. the action part of the rule is evaluated). A very simple but in some cases sufficient solution is to write such specific rules, that only one rule applies in a given context.

Rules interact with WM by adding or deleting items. Those 'internal' actions are the only means for handling flow of control and information between different rules. In other words: rules communicate only via WM. WM contains descriptions of current goals, notes about actions that have been executed and assumptions about the current state of the 'external' world. Therefore, in cognitive models WM is identified with short-term memory. PS rules can be regarded as independent, atomic units of procedural knowledge in long-term memory.

The use of PS for modelling behaviour can be traced back to Newell and Simon's work on human problem-solving (Newell and Simon, 1972). Problem-solving is modelled as a goal-directed activity, which entails finding a sequence of operators leading from an initial state to a final goal. Tasks, as opposed to problems, do not require search in a problem space, but are associated with a fixed solution strategy. The distinction between tasks and problems can thus only be made with respect to a certain human subject.

3 THE GOMS APPROACH TO TASK MODELLING

How is task knowledge represented? The notion of goals, subgoals, methods and operators originally developed in the context of problem-solving can easily be used to represent task-specific knowledge. The GOMS approach developed by Card et al. (1983) is a theory based on this central idea. It uses higher-level descriptions than PS and can be mapped on to PS and thus be made computable (Polson and Kieras, 1985). Let us first have a look at the GOMS theory.

In GOMS, the user's cognitive representation of the task is described by a tree of goals, operators, methods and selection rules. The task decomposition starts with one top-level goal which generates further subgoals according to the hierarchical structure as additional control elements (sequence and selection). These regulate the actual sequencing of goals and operators in the model. Operators are elementary perceptual, cognitive or motor actions which represent the behaviour required for accomplishing the respective goal.

Methods are fixed strategies which consist of subgoals and elementary operators with a predefined control structure. Selection rules are required if alternative goals or methods can be chosen at a certain point.

GOMS models can be used to predict different aspects of user behaviour for a given task. Performance times can be predicted from the number of steps (goals and operators) required to accomplish a task. Method selection can be predicted from selection rules. Also, sequences of user operations can be determined from a simulation based on a GOMS description.

One can make an important distinction between different GOMS models with respect to the grain size of analysis in the models. The granularity of a GOMS analysis is determined by the elementary operators used in the model. Depending on the degree of detailedness which is expressed by the elementary operators, four levels of refinement can be distinguished: At the *unit-task level*, the elementary operator is perform-unit-task. Performance time is simply estimated by the number of unit-tasks the task involves. On the *functional level*, specification and execution of a function (with parameters included) form elementary units. Input of each function and specification of arguments are separate operations on the *argument level*, whereas single keystrokes are operators on the most detailed level of task description (*key-stroke level*).

GOMS models have been successfully applied and tested in the field of text-editing where the sequential processing of single edits can be well expressed in the hierarchic representation of GOMS.

4 COGNITIVE COMPLEXITY THEORY

Based on the GOMS model, Polson and Kieras (1985) have developed a theory of 'cognitive complexity'. The theory is used to analyse and predict the learning and performance of routine skills which occur for instance in the use of text-editors. In the framework of cognitive complexity theory (CCT) a GOMS analysis is transformed into a set of production rules and the production system is used as a simulation model of user behaviour.

Predictions of observable parameters are derived in the following way. Execution time for a given task is estimated by the total number of recognize–act cycles in the simulation. The time to learn a new method is considered to be a linear function of the number of new productions which have to be acquired. Thus transfer of skill between two different tasks is determined by the number of shared productions relative to the total number of productions needed for the new task. This is a modern version of the classical 'common elements' theory of transfer (Thorndike and Woodworth, 1901).

In a number of experimental studies, the predictions from cognitive complexity models could be confirmed (Polson and Kieras, 1985; Polson et al., 1986).

5 COGNITIVE COMPLEXITY TASK MODELS FOR THE EXPERIMENTAL STUDY

The experiment described in this chapter can be regarded as a test of CCT across different application domains. Whereas earlier tests of the theory used either text-editing or (menu-based) operating tasks, in our experiment text and graphics editing tasks were compared. Although the structure of the objects to be manipulated was different for text tasks vs graphics tasks, the same universal commands could be applied in both cases. In terms of a GOMS analysis, there was a functional equivalence of text and graphics editing tasks. On a more detailed level of analysis, selection and cursor positioning turned out to be more complex for text than for graphics.

The functional equivalence of text and graphics editing tasks could only be achieved with a system providing a uniform command structure in both domains. This is a characteristic feature of so-called 'direct manipulation' (DM) systems, which are based on graphical interaction techniques like selection of objects with a mouse.

For our experiment, we used a DM system which makes extensive use of universal commands, thus enabling us to construct similar tasks in both domains. Similarity of tasks is reflected by a high number of shared productions in the CCT model. The transfer predictions (i.e. reduction of learning time for some task B, after having acquired the productions for a similar task A) are therefore very high. The experiment can thus be considered as a strong test of the effectiveness of a consistent user interface design.

Four different types of editing tasks were used, differing with respect to the object domain (text vs graphics) and the aspect being edited (content vs form and layout attributes). The four possible combinations are shown in Figure 2.

Production systems are normally 'flat' and unstructured, which makes them difficult to handle when the rule base becomes large. This was the case for our task model with more than eighty productions. We therefore

	Content editing	Form editing
Text editing	Delete Move Copy	Font Size Feature
Graphics editing	Delete Move Copy	Linewidth Shading Texture

Figure 2.   Task types used in the experiment.

provided our PS interpreter with a mechanism for working with a structured rule base. The rule base consists of rule sets or packages. A new rule set can be invoked by means of a special action. A stack of rule sets is built up and if the current one does no longer respond to the given WM, the previous rule set on the stack is activated (and so forth). From the psychological point of view, this process can be interpreted as setting a focus in a network of methods. More specialized methods are invoked by general ones, and after applying a special method the focus is reset. Figure 3 shows a method graph for all types of tasks used in the experiment. For each method the number of productions involved is indicated.

For a given task and a system, there is not a unique way to formulate a PS model. Even if we exclude alternative strategies of task execution, there are still two degrees of freedom: (1) The grain size of the model, which depends on the terminal actions, and (2) different higher-level task decompositions, which may all produce the same sequence of actions for some given tasks. In this aspect, standardization is necessary to make results comparable. In accordance with other studies based on CCT, we have for our experiment used the following constraints or 'style rules':

— Terminal actions are single keystrokes, perceptual actions or mouse movements (a distinction between long distance and short distance movements is made).
— Every action changing the state of the document (form or content) has to be followed by an explicit perceptual feedback action.

— A rule may not contain more than one external (motor or perceptual) action or one cognitive action.
— No conflict resolution is used.
— Functions and object attributes occurring in productions are not parametrized (whereas objects are).
— The model is strictly deterministic, i.e. there are no alternative methods for any task or subtask.

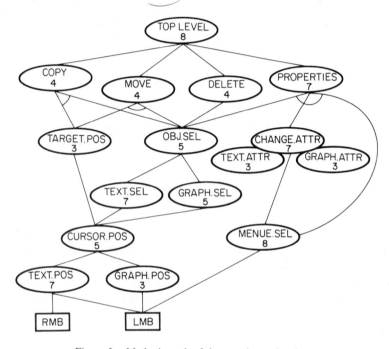

Figure 3.   Method graph of the experimental tasks.

6 PROCEDURE

Four groups of subjects (total $n = 88$) performed the four types of editing tasks mentioned above in four different task orders. Ten instances of each task type were provided, each involving the same number of operations in the same order. There was one session for each task type. Figure 4 shows the task order for each group with the associated number of new production rules to be learned in each session. A session was terminated after three subsequent error-free trials (learning criterion). Learning time was measured as the duration of a session until the completion of the first of these three error-free trials. If the criterion was not reached within a maximum of ten trials, the subject was dropped from the sample.

	S_1	S_2	S_3	S_4
G_1	44 TC	26 TF	9 GC	3 GF
G_2	40 GC	26 GF	13 TC	3 TF
G_3	52 TF	12 GF	18 TC	0 GC
G_4	48 GF	31 TC	3 TF	0 GC

Figure 4. Task orders with numbers of new production rules (G1–4: groups, S1–4: sessions).

The subjects were all novices in order to exclude transfer from previous usage of a computer. The subjects received a minimal standardized introduction by the experimenter. At the beginning of each session more specific instructions were given. During the experiment, help was provided on request or when an error occurred which the subject was not able to correct. Subjects were encouraged to learn exploratively by trial and error.

The material was provided by printed reproductions of the initial and goal states of the document to be edited. In the initial version the necessary operations were indicated by means of colour-coded mark-ups. The duration of the whole experiment was limited to a maximum of four hours for each subject. Four experimenters were randomly assigned to the four groups of subjects.

7 RESULTS

First of all the data were checked for experimenter effects in groups. However, from the analysis of variances there was no indication of experimenter effects, due to the large variability between subjects. Therefore, there was no need to include the experimenter in the statistical model for further analysis. Likewise, no significant effects of sex, age or educational background of the subjects were found.

Figure 5 shows the result of a linear regression analysis applied to the four groups simultaneously. The learning times have been averaged over subjects and regressed upon the number of new productions per session and group. The linear fit between these variables appears to be very good ($r^2 = 0.88$). We found a linear coefficient of 17.2 seconds as an average learning time per production, which is smaller than the estimates found by Polson and Kieras (1985) (c. 30 seconds). A plausible explanation for this difference is that our training procedure involving a human tutor is more efficient than the one used by Polson and Kieras (1985), where subjects had to read unspecific help material presented on a separate screen.

Predicted observations	S_1	S_2	S_3	S_4	Subjects
G_1	955 822	646 636	354 299	251 218	18
G_2	887 791	646 806	423 412	251 287	22
G_3	1093 1143	406 383	509 376	199 252	21
G_4	1024 1280	732 542	251 275	199 304	20
Regression:		$y = 199.5 + 17.2*x,$		$r^2 = 0.88$	

Figure 5.   Predicted vs observed average learning times.

In addition to the strong linear effects described, further significant, non-linear effects could be found which are not explained by the theory. Also, subject variability was very high. These results are not only in accordance with previous findings, they also show that the predictions derived from the similarity of operative knowledge in both task domains are not disturbed by the semantic differences. This confirms the importance of consistent system design. As far as the system used in the experiment is concerned, the expected benefit of a stringent use of universal commands over different domains could indeed be shown.

8 DISCUSSION

The advantages of the application of cognitive modelling techniques to HCI can be summarized as follows:

— Psychological theories of human–computer interaction are turned into exact or 'hard' science (cf. Newell and Card, 1985).
— A causal understanding of user behaviour parameters in relation to the task/system structure is facilitated.
— Cognitive models can guide and, to some extent, even replace empirical work.
— Cognitive modelling techniques can be applied to ergonomic product design in a very early phase of the development.

Cognitive complexity theory is one example of this approach, which has already shown good results in different domains. The study described in this chapter particularly shows the value of CCT for transfer and consistency issues across different task domains. However, there are some inherent limitations to cognitive complexity theory due the fact that the model does not deal with problem-solving, choice of alternatives or errors and regards perceptual actions as undifferentiated 'black box' task units. Also, the different semantics of objects (e.g. types or class relations) are only reflected implicitly in the operative task structure.

Both GOMS and the Polson and Kieras approach deal with fixed strategies of task completion. Features of the system are only considered if they are used for correct completion of the given tasks. Not the whole interaction space offered by the system, but only a path through this space is modelled. There is an underlying assumption that interaction is driven by the user's procedural representation of the task, but this view might not always be adequate. For example, using a menu system the user might find a correct path through the interaction space without any specific procedural knowledge, only guided by the menu choices offered by the system. On the other hand, learning difficulties resulting from semantically misleading menu items could not be explained by a complexity model of correct task completion.

Further developments will be needed, particularly in the following aspects:

— Representation of semantic knowledge about objects and functions. In editing and filing, it is possible to isolate basic procedural skills, which indeed form an important part of the work with the system. In domains like spreadsheet calculation or work with databases, a separation of skills and problem-solving behaviour turns out to be much more difficult.

— Cognitive complexity theory is not a theory of individual differences. Nevertheless, it seems that we need better explanations for the large individual differences. Also, issues of errors and instructions which are not addressed in the current theory will require more attention in the future.

— To make CCT a practical tool for the user interface designer, a higher-level modelling language should be provided. It is a rather clumsy task to write a PS, because all the sequential and hierarchical control knowledge, which is explicit in a GOMS analysis, has to be hidden in the implicit triggering mechanisms of the PS. On the other hand, the PS model reflects the 'cognitive cost' of this control knowledge, thus allowing direct psychological interpretations. A possible solution is to provide the interface designer with a higher level—e.g. GOMS-like language, which can be automatically compiled into a PS.

These problems indicate important directions for research. The representation of semantic knowledge, being a more fundamental issue, could widen the scope of the theory, whereas an adequate modelling language for designers and engineers is of immediate practical importance.

ACKNOWLEDGEMENTS

The statistical analyses have been carried out by Miss S. Evans. We like to thank her as well as the experimenters (B. Mikusin, M. Wöhrle, M. Latzina and J. Herczeg) for their valuable assistance in various stages of our research.

REFERENCES

Anderson, J. R. (1983). *The Architecture of Cognition*. Cambridge, MA, USA: Harvard University Press.

Card, S. K., Moran, T. P. and Newell, A. (1983). *The Psychology of Human–Computer Interaction*. Hillsdale, NJ, USA: Erlbaum.

Fähnrich, K. P. and Ziegler, J. E. (1984). Workstations using direct manipulation as interaction mode. In: *Proceedings of Interact 1984*, London 4–7 Sept. '84, Vol. II, pp. 203–8.

Hutchins, E. L., Hollan, J. D. and Norman, D. A. (1986). Direct manipulation interfaces. In: *User Centered System Design—New Perspectives in Human–Machine Interaction*. D. A. Norman and S. W. Draper (eds). Hillsdale, NJ, USA: Erlbaum.

Johnson-Laird, P. N. (1983). *Mental Models.* Cambridge, UK: Cambridge University Press.
Newell, A. and Card, S. K. (1985). The prospects for psychological science in human–computer interaction. *Human–Computer Interaction*, 1/85, 209–42.
Newell, A. and Simon, H. A. (1972). *Human Problem Solving.* Englewood Cliffs NJ, USA: Prentice-Hall.
Polson, P. G. and Kieras, D. (1985). An approach to the formal analysis of user complexity. *International Journal of Man–Machine Studies*, 22, 365–94.
Polson, P. G., Muncher, E. and Engelbeck, G. (1986). A test of a common elements theory of transfer. *Proceedings of CHI '86*, April 1986, Boston (MA), pp. 78–83.
Shneiderman, B. (1982). The future of interactive systems and the emergence of direct manipulation. *Behaviour and Information Technology*, 1(3), 237–56.
Smith, D. C. et al. (1983). Designing the star user interface. In: *Integrated Interactive Computer Systems.* P. Degano and E. Sandewall (eds). Amsterdam: North Holland.
Thorndike, E. L. and Woodworth, R. S. (1901). The influence of improvement in one mental function upon the efficiency of other functions. *Psychology Review*, 8, 247–61.
Williams, G. (1983). The LISA computer system. *Byte*, 8, 33–50.

THE ROLE OF MEMORY IN PERSONAL INFORMATION MANAGEMENT

Mark LANSDALE

Cognitive Ergonomics Research Group, Department of
Human Sciences, University of Technology,
Loughborough, Leicestershire UK

1 INTRODUCTION

We all keep information in our work. It may be books, notes, folders, diaries, or whatever. This is personal information not only in the sense that it is private, but that we keep it for our own use. This chapter considers some of the psychological processes involved in the management of this information: the methods and procedures by which we handle, categorize and retrieve information on a day-to-day basis. This represents an area in which psychological theory can contribute to the satisfaction of a clear technological need. We are witnessing in the emergence of Information Technology (IT) a facility in which there is a pressure to create, store and process more and more information. But the purpose of IT should be to increase the quality, not merely the quantity, of available information. If all it achieves is to increase the volume of our filed information without an associated improvement in retrievability, then the reverse must be expected.

The psychological aspect of this is that the management of personal information, by definition, involves psychological processes. The information to be retrieved has already been handled, categorized and filed away by that individual. It is therefore reasonable to ask how much the problems of information management stem from these processes and the ability to remember what was done. Consider some examples:

'My boss wants to see all the project reviews I have carried out over the last six months. The trouble is, they are filed under each of the individual projects. It will take me ages to work through and dig them all out.'

'I know what the thing looked like: it had a blue and white stripe at the top, but I can't remember how I filed it, or even what it was about.'

COGNITIVE ERGONOMICS:
UNDERSTANDING, LEARNING AND DESIGNING
HUMAN–COMPUTER INTERACTION

'Yes I remember that paper. It came at the same time as the product audit. I can't remember what happened to it though.'

'The document I want is the French Finance Committee's minutes, but I've tried looking under "French", "Finance", and "Committees", and it's not there. Perhaps its under "France". I may as well search through the lot.'

I would not like to suggest that all the problems of information management are psychological in nature. But these examples illustrate two important issues in the development of information management tools on computers. First is that there is a general problem in categorizing items (e.g. see Furnas et al., 1983), and second is the assumption (as yet formally untested, but see Lansdale, 1988) that we remember far more about the documents than can be used in retrieval procedures. Clearly, if the first of these could be ameliorated, and the second exploited, powerful tools will emerge. This chapter attempts to articulate a research policy aimed towards producing such tools.

2 STUDIES OF THE 'NATURAL HISTORY' OF OFFICES

One approach to the psychology of information management is to study the natural history of offices (e.g. Cole, 1982; Malone, 1983). The notion is that by studying the behaviour of people in natural situations we may come to understand the underlying psychological principles. The small study by Malone is a good example of this. Ten office workers were interviewed about their jobs, their information management habits and their ability to access target documents. The dispositions of their offices in terms of layout, distribution of papers, filing methods, etc., were also examined.

A principal distinction that Malone draws is between 'neat' and 'messy' offices. We should not be surprised to find out that occupants of neat offices reported fewer difficulties in information retrieval, overlooked fewer things they had to do, and were better able to find specified target documents on request. Clearly some of this is personal style. Some people are tidier than others. However, this may not be the whole story. Malone's analysis suggests that the 'mess' in offices is not arbitrary, but reflects a response on the part of the individual to particular needs or difficulties. Three in particular are identified. First, documents are left lying around in conspicuous places as reminders that something has to be done with them. Second, some tasks need information from several sources, and therefore require that several documents are kept close and 'active' while the task is being carried out. Third, Malone suggests that on many occasions people are reluctant to file information away either because they cannot decide how to categorize it, or because they are not confident in their ability to retrieve it later. As a result, the compensating strategy is to pile documents around the office in relatively unstructured files.

If one accepts this view, people do not allow their offices to become messy purely because they are slovenly, although this may well be a contributory factor. There is also a mismatch between what the person needs to do and the facility to do it. All these things create the pressure to spread documents around the office in more or less unstructured piles. The 'technology' (in this case paper-based) does not adequately support these needs, and ultimately the strategy of using piles of information is counter-productive.

3 PROBLEMS IN THE INTERPRETATION OF OFFICE BEHAVIOUR

There are interesting things we can learn from observing natural information management behaviour. No doubt more detailed systematic studies will reveal more. However, there is also a note of caution required. It might seem straightforward to suppose that we can translate observed strategies of information handling from existing paper-based methods to computers. This is, as we shall see later, the basic philosophy of many leading producers of office automation equipment such as Xerox (Smith et al., 1982). As a design principle, however, this must be a mistake. Consider the use of piles identified in Malone's study. No-one would suggest the introduction of unstructured 'piles' of documents in a computer environment. Apart from anything else, they are evidently counter-productive. How then do we decide which aspects of office behaviour to emulate in office automation and which to avoid?

The problem is that the strategies used by people in one technology need not apply to another. The point is subtle but critical to a psychological analysis of information management. The piles that Malone reports are not, in a simple sense, representative of a NEED in the user. Quite the reverse, in fact. They are a compensating strategy for the problems of classification (e.g. see Dumais and Landauer, 1983). In using piles, office workers are making a trade-off along several dimensions of difficulty. To avoid the process of classification they put objects in a particular place. With this they forgo the opportunity to retrieve the document by any simple classification-based search. But they have other things up their sleeve. First they may remember what it looked like; consequently a strategy of scanning the piles (they are of course open to inspection) may identify the target by visual recognition. Second, they may remember where it was left, thereby reducing the area for search. Third, the use of piles has an implicit element of ordering by time: the most recent documents are near the top of the piles. This may also be useful information.

As an aside, it is worth emphasizing the word 'may' at the end of the last paragraph. Which of these strategies, and others as yet unspecified, are actually used by people in offices is an empirical question to be resolved. At this stage, the important thing is to recognize that existing offices offer the opportunity for a diverse range of compensatory strategies. As we will see, it is the flexibility this affords which we must aim to build into future systems, and which is missing in many contemporary products.

Returning to the point at hand, concentrating upon piles is to caricature what happens when procedures from office practices are reapplied to computers, but it illustrates the point well. Merely to invoke the concept of PILES in such systems without an understanding of the role played by recognition memory, or memory for the spatial and temporal attributes of documents, may be entirely to miss the point: the piles may be the visible manifestation of an information handling strategy, but the psychology that underlies them may be much less direct. Only an understanding of this psychology can lead to genuine advances in information retrieval systems. To illustrate this point, I now turn to the main focus of this research: how an understanding of human memory, as exploited in current filing systems, can lead to useful design principles for the future.

4 HUMAN MEMORY AND FILING SYSTEMS: A FRAMEWORK FOR FUTURE RESEARCH

The framework I propose for the process of information retrieval is very simple: it is that EVERY attempt at retrieving information involves two distinct psychological processes: recall-directed search, followed by recognition-based scanning. Recall-directed search refers to the use of memory about the required item to get as close to the document as possible. At its most exact, this will give direct access, such as remembering a filename or in which sleeve of a filing cabinet drawer it is located. As it becomes less precise, memory might identify an area of the database: a drawer of a filing cabinet, a shelf, or perhaps a particular computer directory. In this case, more or less recognition-based scanning within that area is required, depending upon the exactitude of recall. Recognition-based scanning is therefore the process we undertake when recall has failed to produce a unique item, and the amount we have to undertake depends upon how specific the recall was. This is, of course, directly analogous to direct-access and inferential methods of access from human long-term memory (e.g. see Herrmann, 1985, p. 263).

Accordingly, to study the psychological issues addressed, I find it useful to propose that the observed behaviour with any information storage and retrieval system be interpreted in terms of a trade-off between these two processes of recall and recognition. Users can be said to be balancing the two processes to maintain, in their eyes, reasonable utility. In other words, the strategies people adopt can be seen as a way of shifting this trade-off to their advantage. For example, by leaving documents around an office in piles, people are using their recall for where it was placed, when it was placed there, and what it looked like. The scanning process can therefore be reasonably efficient because recall can not only specify where to look, but also visual attributes which help in recognizing the target.

Continuing this example also illustrates where new technology can improve on existing practices when users' strategic thinking does not extend beyond the short term satisfying of immediate problems. Thus, as piles of information accumulate, scanning becomes less and less efficient. If new computer-based tools supported scanning techniques without forcing users into long-term difficulties, then we get the benefits of the old strategies without their disadvantages.

The balance between recall and recognition is also a way of looking at the usability of existing systems. For example, it can be seen why traditional operating systems running on the basis of using single filenames fail to provide an adequate solution to the storage and retrieval needs of the electronic office. The recall process is limited by the user's ability to categorize documents with the appropriate filenames, and their ability to remember those filenames (e.g. Furnas et al., 1983). On the other hand, in these systems the user cannot easily fall back upon scanning methods because it is enormously clumsy to search directories and open files one at a time to inspect their contents. The user simply has nowhere to go; with no way of using the machine which will provide an acceptable trade-off between the problems of recall and recognition. On the other end of the scale, the Spatial Data Management System (Bolt, 1979) relies exclusively upon recognition-scanning strategies, and although unevaluated, it is hard to see how this could provide realistic support for the management of real information systems.

Thinking in terms of this trade-off also suggests a general method of approaching the design of an information management system. A future system should be as near to an optimization of the two processes as possible: recall processes should allow the users to use whatever memory they have to limit the area of the database to be searched; and then the information within this area of the database should be represented in such a way as to optimize the search process. I see no evidence to suggest that those methods which support recall should have any detrimental effects upon recognition, or vice versa. Therefore, optimizing both processes (as far as is possible) should give any user enough room to manoeuvre such that they can accommodate their special needs and idiosyncracies.

Approaching the design of personal filing systems as an issue in human memory is not an entirely novel philosophy. For example, from a rather similar theoretical position, Jones (1986) has been developing a so-called 'Memory-Extender' filing system. Thus far, however, it has concentrated upon textual information as a prompt for information search, and has been concerned with applying a specified model of human memory to the problem. This is clearly important, but our philosophy of research is directed at the flexibility of human information processing strategies, both in terms of the methods adopted and the modalities of the information handled. That is, we are less concerned with finding a particular model around which to organize a filing system, than with understanding how to deliver a range of methods which will provide the kind of flexibility currently available in offices, and some more functionality beyond that. This leads to a number of specific research issues we have been concerned with in the laboratory.

5   EXPERIMENT 1: WHAT IDENTIFYING ATTRIBUTES SHOULD ELECTRONIC DOCUMENTS HAVE?

Electronic systems introduce new degrees of freedom in the possibilities of information management. We are no longer constrained to see documents as strings of text with a name, and we are entitled to ask how all sorts of additional attributes such as icons, colours or pseudospatial metaphors (e.g. see Card and Henderson, 1987) can be used to create a more rich and varied environment which exploits aspects of human memory. This is the philosophy of 'cue-enrichment' (e.g. see Cole, 1982). Some research has been carried out upon the memorability of different methods of coding documents (e.g. Wright and Lickorish, 1988), but little is associated directly with information retrieval. For this reason, an exploratory experiment was devised in which subjects were required to code documents by associating a shape to them whose colour and location on the document were also at the user's discretion. The coding method is therefore one of creating a coloured 'watermark' on the document which incorporated three dimensions. This was chosen because strong claims are being made for the memorability of icons (e.g. Lodding, 1983), and because they provide a convenient way of combining a number of visual attributes such as shape, colour and location. Each of these is of mnemonic interest in its own right.

This experiment is documented in more detail in Lansdale et al. (1987) and is described only in its essential details here. The users were provided with a simulated system in which they were asked to imagine they were working in an employment agency. They were then shown a sequence of eighteen job adverts which they were required to file in the following way. First, a selection of the required shape was made from a touch-sensitive array of twelve shapes, chosen such that six of them had some meaningful relationship with the adverts being presented. Following this, a selection of colour from one of twelve was made in much the same way, and finally, the user pointed to one of twelve locations on the advert to place the coloured icon. After a delay, memory was tested by re-presenting the adverts to the users (in a different sequence) and asking them to reconstruct the original icon in the same way as before. For comparison purposes, a second group of subjects was asked to carry out the same task using a verbally based system in which they assigned three names, a category, colour name, and adjective, to the adverts. These were also chosen from touch-sensitive menus such that the only differences between the iconic and verbal methods were the mode the file-designators were coded in, iconic or verbal.

The key results of this experiment were as follows. First, no simple picture of the benefits of iconic as opposed to verbal methods emerged. The users appear to perform at roughly the same level for both systems. More important to recall appears to be the degree of semantic link between the attributes associated and the adverts. Thus, the icon shapes and the category words are recalled well, colour and colour names less well, while adjectives and locations were recalled relatively poorly. This latter result confirms a study by Dumais and Jones (1985), that memory for arbitrary locations is not particularly effective. Further analysis of the coding strategies of the users implies that those attributes which were less well remembered were harder to code when assigning them to adverts. In any case, it is not clear that iconic methods are the panacea to recall that might have been expected (e.g. see Lodding, 1983).

A second important result of this experiment (also found in Lansdale, 1988) was that recall of the coding dimensions was statistically independent: the likelihood of recalling colour, for example, was not contingent upon whether the user had already recalled shape or location. The characteristic of an independent model of recall is that users' memory for particular icons or word combinations will be predominantly partial. It is unlikely that nothing will be recalled at all or that all of the combination will be recalled. This suggests that an information management tool which allowed users to employ their partial knowledge will be of considerable utility. Lansdale (1988) developed such a tool which increased the accuracy of access by an order of magnitude. This tool is currently the focus of ongoing research.

6 EXPERIMENT 2: HOW EASY SHOULD FILING BE?

One aspect of information retrieval which is important to emphasize is that although difficulties may be apparent at the retrieval stage, in that one might not be able to find something, it is as sensible to see this as a problem of storage as of retrieval. We cannot find information because we did not file it in a form which is appropriate for future retrieval, or perhaps we didn't pay enough attention to it. It is a natural reaction for people not to spend a great deal of time filing information because it has no immediate pay-back and because they want to get on with the next piece of work. As might be expected, a parallel result is to be found in the retrieval of information from human memory, where organization of material at the time of learning improves its recall (e.g. Bower et al., 1969).

A second condition in Lansdale et al. (1987) shows that users assigning retrieval tags to documents (pictures and words) have much stronger memory for those tags if they assigned them themselves as opposed to having them assigned by the system. This was demonstrated by presenting to the subjects the adverts with the icons or file words already assigned by a different subject. Thus, the only difference between subjects in this experiment and the previous one is that they have no control over the assignation of the coding. This result ties in well with well-established educational principles that memory is much more robust for self-generated material than when it has been provided for the subject (e.g. Bower, 1970).

This leaves us with a design dilemma which is also the focus of ongoing research. The more we ask the users to do at the process of information storage, the less likely they are to do it, creating retrieval problems. On the other hand, the more we automate the process of storage and take responsibility away from the users, the less they are going to remember, and therefore the less they are going to be able to retrieve. This leads to two classes of research issues which follow. First, we can ask how we might facilitate necessary tasks such that they are not so onerous that users will not carry them out. This is a matter of finding out what it is we want the user to do at the stage of storage and considering how the software could be devised to do this most easily. Alternatively, we can resign ourselves to automated systems in which the user's memory is intrinsically poorer. In this case we are interested to ask questions as to what, if anything, is best remembered under these conditions and how we can best exploit what is remembered. The key point here is that any future research must be sensitive to the trade-off between the positive aspects of easy filing facilities and the negative effects they have upon the user's memory.

7 SUMMARY

One aim of this chapter is to provide some justification for the view that the study of human memory has applied value in the design of personal information management systems. This is essentially based upon the following premises:

1. Paper-based office systems distribute their information around a wide variety of items and incorporate, either explicitly or implicitly, a diverse range of dimensions such as location, visual appearances, and so forth.
2. These enriching dimensions may confer psychological advantages in their use. This remains to be established reliably, but we can expect that this is more likely to be true in particular situations, for which these dimensions are appropriate. This means, as the point below emphasizes, that we should not expect new, singular, methods to provide general improvements in usability.
3. Notwithstanding the values of these enriching dimensions per se, their diverseness in existing offices allows for great strategic flexibility for their user. This provides a significant advantage over computer-based systems which are, for the most part, restrictive in the strategies open to the user.

The research approach we are currently adopting as a means of testing these premises, and turning what we discover into application, is broadly four-pronged:

1. Observational studies are under way to identify common effective strategies in office information management.
2. Such strategies as are identified are respecified in terms of their underlying psychology. If their purpose is positive (that is to say, they perform a useful filing function as opposed to merely compensating for some shortfall in the existing technology) then we attempt to use computer-based technologies to fulfil the same function.
3. We recognize that computers provide for some strategies not possible in exsisting offices. Insofar as they are identifiable, we attempt to specify these also.

4. Tools (computer-based methods designed to support the identified information management strategies) are incorporated into a computer-based
system we are currently developing. When completed, we will use this
system for a period of at least two years as an information system kept in
parallel with our everyday offices. The comparison of the two systems in
use over these periods of time provides the basis of the evaluation of our
premises. It will also, of course, tell us a great deal about many other
issues involved in the design of information management systems.

To many, this 'suck it and see' approach may seem a somewhat inelegant
strategy of research. By way of justification, I return to the basic premises,
because they dictate this approach. If we accept that human beings are
strongly strategic in their behaviour, then controlled experiments of the kind
I have described must be limited in what they can usefully tell us about the
real world, and therefore how to design systems (e.g. see Neisser, 1978 or
Newell, 1973). This is because the subject's response to any particular
experiment is strategic, and will reflect a flexible and possibly ungeneral-
izable response to the particular task demands. On the other hand, when it is
those strategies which are themselves of interest (for example, we might want
to know the circumstances under which one filing strategy is preferred over
another), then we can only observe them by creating the task in which they
are observed. When we are talking about personal information management, this means everyday information management, over realistic periods
of time, and over realistic quantities of information. Similar comments have
been made about the study of autobiographical memory, to which this
research must be closely related (Linton, 1978; Wagenaar, 1986), and I see
the slow uptake of these approaches as grudging acceptance of the fact that
elegant experiments do not always lead to useful insights or, in an applied
arena, to good design.

ACKNOWLEDGEMENTS

This chapter is based upon a paper given at the meeting of ECCE3, Paris,
1986. The author is grateful to Debbie Young for her useful comments on
that paper which have contributed to this version.

REFERENCES

Bolt, R. A. (1979). *Spatial Data Management*. Boston, USA: MIT Press.
Bower, G. H. (1970). Analysis of a mnemonic device. *American Scientist*, 58, 496–510.

Bower, G. H., Clark, M. C., Lesgold, A. M. and Winzenz, D. (1969). Hierarchical retrieval schemes in recall of categorized word lists. *Journal of Verbal Learning and Verbal Behaviour*, 8, 323–43.

Card, S. K. and Henderson, A. (1987). A multiple, virtual-workspace interface to support user task switching. In: Proceedings of CHI + GI '87. Special Issue of *SIGCHI Bulletin*. J. P. Carroll and P. P. Tanner (eds), pp. 53–60.

Cole, I. (1982). Human aspects of office filing: implications for the electric office. *Proceedings 26th Annual Meeting of the Human Factors Society*, 25–29 October, 1982, Seattle USA.

Dumais, S. T. and Landauer, T. K. (1983). Using examples to describe categories. *Proceedings of CHI '83*, Boston, USA.

Dumais, S. T. and Jones, W. P. (1985). A comparison of symbolic and spatial filing. *Proceedings of CHI '85*, San Francisco, USA.

Furnas, G. W., Landauer, T. K., Gomez, L. M. and Dumais, S. T. (1983). Statistical semantics: analysis of the potential performance of keyword systems. *Bell System Technical Journal*, 62, 1753–806.

Herrmann, D. J. (1985). Remembering past experiences: theoretical perspectives past and present. In: *New Directions in Cognitive Science*. T. M. Shlechter and M. P. Toglia (eds). Norwood, NJ, USA: Ablex.

Jones, W. P. (1986). On the applied use of human memory models: the memory extender personal filing system. *International Journal of Man–Machine Studies*, 25, 191–228.

Lansdale, M. W. (1988). On the memorability of icons in an information retrieval task. *Behaviour and Information Technology*, 7(2): 131–151.

Lansdale, M. W., Simpson, M. and Stroud, T. R. M. (1987). Comparing words and icons as cue enrichers in an information retrieval task. *Human–Computer Interaction–INTERACT '87*. H.-J. Bullinger and B. Shackel (eds). Amsterdam: Elsevier Science Publishers BV (North-Holland).

Linton, M. (1978). Real-world memory after six years: an in vivo study of very long-term memory. In : *Practical Aspects of Memory*, M. Gruneberg, P. Morris and R. N. Sykes (eds). New York: Academic Press.

Lodding, K. N. (1983). Iconic interfacing. *IEEE Computer graphics and applications*, 3, 11–20.

Malone, T. W. (1983). How do people organise their desks? Implications for the design of office information systems. *ACM Transactions on Office Information Systems*, 1(1), 99–112.

Neisser, U. (1978). Memory: what are the important questions? In: *Practical Aspects of Memory*, M. Gruneberg, P. Morris and R. N. Sykes (eds). New York: Academic Press.

Newell, A. (1973). You can't play 20 questions with nature and win. In: *Visual Information Processing*. W. G. Chase (ed.). New York: Academic Press.

Smith, D. C., Irby, C., Kimball, R. and Verplank, W. (1982). Designing the star user interface. *Byte*, April 1982, 242–82.

Wagenaar, W. A. (1986). My memory: a study of autobiographical memory over six years. *Cognitive Psychology*, 18, 225–52.

Wright, P. and Lickorish, A. (1988). Colour cues as location aids in lengthy texts on screen and paper. *Behaviour and Information Technology*, 7(1), 11–30.

HUMAN–COMPUTER INTERACTION: LESSONS FROM HUMAN–HUMAN COMMUNICATION

Pierre FALZON

INRIA, Rocquencourt, BP 105, 78150, Le Chesnay, France

This chapter is dedicated to the lessons that can be learnt from studies in human–human communication to improve human–computer interaction. The first question to consider is the applicability (or relevance) of the human–human communication model to human–computer interaction. The following questions will then be addressed:

— What type of language is used?
— What interaction mode is used?
— What dialogue strategies are used?

Since written (or typed) dialogue is not natural in human–human communication, I focus on speech communication (although occasionally I refer to computer-mediated communication).

1 THE APPLICABILITY OF THE HUMAN–HUMAN COMMUNICATION MODEL

A number of studies have examined the differences between human–human and human–machine voice dialogues. These studies use variations of a single experimental paradigm. Each subject has to communicate with an (invisible) interlocutor, in a more or less complex situation of information query. For one group of subjects, the interlocutor is presented as being a human; for the other group, it is presented as a machine. The machine is in fact simulated by an experimenter (Wizard of Oz method). Variations between studies concern the quality of the voice heard by the subject and the degree of constraint on the experimenter's production and comprehension.

The results presented below are compiled from the conclusions of Chin (1984), Richards and Underwood (1984), Morel (1985), Spérandio and

COGNITIVE ERGONOMICS:
UNDERSTANDING, LEARNING AND DESIGNING
HUMAN–COMPUTER INTERACTION

Létang-Figeac (1986), and Meunier and Morel (1987). These studies indicate that, in the human–machine condition, human verbal production is modified in the following ways:

- Attempts to simplify the communication:
— decrease of the speech rate and volume
— decrease of the conversation length
— increase in the number of dialogue interventions
— decrease in intervention linking (more moments of silence, less interruptions, less overlapping)
— less comments and digressions
— increase in explicit planning ('First, I want . . .')
— increase of the double questions ('I want this and that')

- Attempts to simplify the expression:
— less contextual expressions (anaphors, ellipses, pronouns)
— smaller vocabulary
— preferential use of some syntactic structures
— reduction of the verbal group

- Attempts to 'normativize' the language (i.e. production of a language closer to written language):
— less indirect requests
— less uncompleted sentences
— more well-formed sentences
— less restarts

- Less dialogue structuration:
— less linking expressions (well, I mean, . . .)
— less markers
— neutralization of markings (with 'and')
— less expressions that link the themes

- Less dialogue control acts:
— less readbacks, less acknowledgements
— less expressions of politeness
— less expressions that close the communication channel
— less expressions that maintain the communication channel

However, it can be argued that some of these results are engendered by the characteristics of the (simulated) machine (e.g. distorted voice, restrictions in production or comprehension) and not only by the fact of speaking to a machine. In these studies, experimental conditions are not similar for the subjects interacting with the machine and for the subjects interacting with a human being. In an experiment conducted by Amalberti et al., 1986, the linguistic characteristics of the user's interlocutor (human or machine) were strictly identical (the experimenter did not know the role she was playing). The observations contradict many of the results just listed:

— There is very little difference between the two conditions (human–human vs human–machine). Thus, when humans modify their verbal productions, it is not so much because they speak to a machine, but because the machine appears to have limited abilities.
— Even when subjects modify their language, it does not always result in a simplification. For example, data indicate that subjects tend to use a larger vocabulary when addressing the machine.

Verbal productions depend very much on a model of the interlocutor abilities. When the interlocutor is a machine, this model is at first based on the prior experience with other machines, and then derived from interacting with the machine.

Concerning the role of prior experience, Onorato and Schvaneveldt (1986) have studied the task of specifying procedures to different partners, including a human partner and a computer. They conclude that the relevance of the human–human model is not independent of the experience of the subjects: users differing in experience with computers have different expectations about what the computer is capable of doing. Totally naive users tend to behave with computers as they do with human beings. Beginners and experts behave differently.

Concerning the elaboration of a model of the interlocutor, the experiment of Amelberti et al. (1986) shows that, if the machine abilities equal those of a human, there is little difference in the linguistic behaviour of the subjects. The subjects' behaviour is a consequence of the behaviour of the machine. A similar result has been reported by Zoltan-Ford (1984). She has shown that it is possible to modify some aspects of the subjects' behaviour by adapting the behaviour of the machine.

The problem is then to provoke appropriate modifications of the users' linguistic behaviour. Modifications are appropriate if they result in an acceptable language for the machine. This means defining the machine linguistic behaviour that allows the user to build an accurate model of the machine abilities.

2 THE LANGUAGE AND THE MODE OF INTERACTION

2.1 Is Natural Language Natural?

I begin with a contribution to a long-lasting debate concerning the language that should be used in human–computer interaction. Two positions have been held, the first advocating natural language, the second in favour of restricted command languages.

Several experimental results cast doubts on the use of natural language in human–computer dialogues:

— given the choice between natural language and a restricted language, users often choose the latter (Gould et al., 1976; Hauptmann and Green, 1983);
— even in the absence of all contraints, subjects tend to restrict spontaneously the language they use (Scapin, 1986);
— natural language does not allow a better performance than a well-designed restricted language (Hendler and Michaelis, 1983; Bailey, 1985; Borenstein, 1986).

However, two objections can be raised. First, many studies have concerned typed interaction, considering that the state of the art in speech comprehension forbade the possibility for oral communication with machines. One can argue that different results would have been found, had oral communication been considered. The preference for restricted languages, observed in typed interactions, has its origins, at least in part, in the constraints of keyboard inputs. This effect is of course stronger if subjects are not experienced typists (cf. Chapanis, 1978).

Second, it appears that humans have a natural tendency to restrict their expression, not only when typing instructions to a machine, but also when interacting with other humans. This tendency, described by many authors (e.g. concerning the process of referring, the studies by Krauss and Weinheimer (1964), Krauss and Glucksberg (1977) and Clark and Wilkes-Gibbs (1986) may even lead, especially in work contexts, to the elaboration of specialized sublanguages, operative languages, restricted and distorted as compared to natural language (Falzon, 1983, 1984, 1989, in press). Operative languages have the following characteristics:

— their vocabularies are limited in size, and are neither a representative sample nor a subset of the total vocabulary of the language: they include words which are rare or unknown in the language as a whole;

— their syntax is limited (their grammars have a limited set of rules), and specific in two ways: some of the syntactic rules of operative languages do not exist in the general grammar of the language, each operative language has its own grammar;

— the words of operative languages tend to be monosemous. The meanings of the words may be rare or unknown in the language as a whole. The same word, if used in several operative languages, will have a specific meaning in each operative language. Moreover, semantic restrictions tend to limit potential ambiguities;

— pragmatic interpretations are limited by restrictions on the set of possible goals of both dialogue partners, and by a tighter application of conversational rules.

That humans elaborate operative languages does not mean that designing natural language understanding systems is useless, but certainly indicates that the decision to use natural language in computer interaction is not an obvious one, and that restricted command languages are not as unnatural a solution as some claim it to be.

2.2   Interaction Modes

Here again, we face a well-known question, that is: What is the optimal mode for man–machine communication: menu vs form fill-in vs command language vs natural language, etc. Experimental results tend to indicate that the choice of a mode depends upon the users' experience: the menu is better suited to novices, the command language is better adapted to the experienced operators (or to those who are to become experienced operators). In other words, experiments seem to demonstrate that, for dialogues with non-experts in the domain, machine guidance is more efficient than user guidance (Kupka and Wisling, 1980; Benbasat et al. 1981; Ogden and Boyle, 1982; Hauptmann and Green, 1983; Ramsey and Grimes, 1983; Potosnak, 1984).

Studies of human dialogues certainly indicate variations of the interactions according to the level of knowledge of the dialogue partner (e.g. Krauss and Glucksberg, 1977). However, I will here emphasize another factor of variation: the goals of the speaker. As will be seen, this is not independent of the choice of an interaction mode. Two dialogue situations can be differentiated: those involving two experts in a domain and those involving an expert and someone seeking some advice, information, etc. These two types of situations call for different dialogue strategies.

2.2.1   Expert–information seeker dialogues

Conversations between doctors and patients are examples of expert–information seeker dialogues. In these situations, doctors can be considered not only as specialists of a domain (medicine), but also as specialists in a specific situation of communication, the medical interview.

These conversations have been studied by Evans (1976). The goal of this author was the automatization of some medical interviews and the conception of an expert system for medical diagnosis which could be consulted by the patients themselves (and not only by the specialists, as it is the case generally). In this perspective, Evans analyses the strategies used by the doctors during the interview, and the patients' answers: the complexity of these answers determines the complexity of the understanding system to be designed. Evans' conclusions are:

— although physicians experience much trouble in trying to describe the procedures they use in their interviews, they do use a dialogue strategy, that can be formalized by rather simple flowcharts;
— the patients' answers use a very restricted vocabulary, and are often limited to 'yes', 'no', 'don't know' answers. Even when answers are longer, physicians always code them as simple statements. When answers get verbose or inconsequential, physicians ask for a 'yes or no' answer.

These results are interesting in terms of system design. Evans even got to the point of designing a keyboard with three keys: 'yes', 'no' and '?' (evaluations of the system are presented in Bevan et al., 1981, and Card and Lucas, 1981). These results are also interesting in terms of the physicians' cognitive processes. For the physicians, comprehension means getting answers that allow them to organize the patient's symptoms, and to classify the illness in a known category. Physicians are thus looking for answers that allow choices to be made between alternatives. In this purpose, they take control of the dialogue as soon as possible, and seek one of the three types of answers described, even if, to get this answer, they have to translate the patient's utterance or to force one of these answers when translation is impossible.

The results presented earlier concerning human–machine dialogue situations seem thus in agreement with Evans' observations on medical interviews: guidance by the expert (physician or machine) is recommended when the dialogue partner is a non-expert (patient or domain-naive subject).

2.2.2   Expert–expert dialogues

Nevertheless, it must be stressed that Evans' results are interesting only in as much as we deal with the consultation of an expert system by domain-naive subjects. Supposing that experts (e.g. physicians) consult the same system, they probably would not be satisfied by it. When human experts consult an expert system, their goals differ from the goals of a patient. Human experts want to check their diagnosis. Of course they are interested in the result proposed by the system, but, more important, they are interested in the reasoning process that yielded this result.

The results given by Clancey (1983) are consistent with these remarks. Clancey has worked on the expert system MYCIN, with the goal of extending the system's explanation capabilities, to make it usable for teaching purposes. In this perspective, although the prospective users (the students) are not experts in the domain, their goals are very similar to those of the experts. For example, they are interested as much in the processing logic and in the justification of the rules as in the diagnosis itself. The analysis of MYCIN by the author is thus focused on the rules and their justifications. Clancey describes different shortcomings in the existing explanation system, caused mainly by the lack of hierarchy among the rules and by the absence of some levels of explanation.

Consequently, the dialogue system necessary for the consultation of an expert system by an expert will differ from the dialogue system necessary for the consultation of that same system by a naive subject. The expert–expert dialogue system must be able to give explanations about its own functioning, about the rules underlying its choices, about the reasoning steps, etc.

3 COMMUNICATION COMPETENCE AND PROCESSING STRATEGIES

When communicating, humans manage to lower the level of complexity of cognitive processing, by using different strategies. These paragraphs will consider different (human) ways of coping with complexity, and their applicability to human–computer interaction.

3.1 Communication Competence

The following remarks deal mostly with expert–information seeker dialogues. In these dialogues, professional competence can be divided into:

— domain competence, i.e. the technical knowledge of a profession;
— communication competence, i.e. expertise in dialogue procedures.

Roman Parré (1984), who has analysed various situations of professional dialogues, suggests that, while technical competence is of course domain-dependent and shared with operators whose task does not include communication with non-experts, this is not the case for communication competence. This second level of competence is independent of the domain and is common to operators in charge of answering queries, whatever the technical domain. For example, income tax advisers in charge of answering telephone calls have the same theoretical knowledge as their colleagues at the same level of the hierarchy, but their 'know-how', e.g. the techniques they use during query answering are similar to those used by other 'query answering' specialists studied by Roman Parré.

Communication competence is the result of a learning process. In a study of telephone operators of a medical centre (Falzon et al., 1986), two operators equal in domain experience and unequal in experience in communication were used. The operator having less experience in communication proceeded differently compared with the more experienced operator. Some of the operators studied by Roman Parré are aware that this competence has been acquired, and are able to spot some of the errors that experience has allowed them to eliminate. For example, operators have discovered that they waited for too long before speaking; they have learnt to interrupt the users' discourse with yes or no questions. Expert communicators take control of the dialogue as early as possible, to save time: experts know the relevant questions, those which are necessary to classify quickly the requests.

3.2 Processing Strategies

Most present understanding systems are characterized by a single mode of functioning: they try to analyse the inputs by using all the time all the available procedures and knowledge structures. The following observations indicate another possible organization, in which systems would be able to use two different modes of functioning.

3.2.1 Simplifying by assuming a correct problem space

In many work situations, a single operator processes different kinds of problems, whatever the level of complexity. This makes it difficult to point out the variations in cognitive processing according to problem complexity. This analysis is easier when problems of different levels of complexity are allotted to different operators. Such a situation is described by Roman Parré (1984), in a study of the task of telephone operators in charge of providing information on telephone numbers. Two categories of operators are involved, each category being allotted a specific task. Operators of the first category receive all calls, and are not allowed more than three attempts for each call. If these attempts fail, the call has to be transferred to an operator of the second category, in charge of 'special queries'.

The underlying assumption of this organization is that two classes of situations occur. First, situations where the request can be processed directly, assuming its formulation to be correct: the task consists only in translating the request into a query to a database (the telephone directory). Second, situations where the call is more difficult, and where, before querying the database, the operator has to analyse the request more fully. 'Standard' cases are then processed by the first category of operators, more unusual cases by the second category. Thus, each category of operators is allotted a specific mode of processing: if the routine mode fails, calls are transferred to the operators in charge of developing more elaborated search strategies.

Of course, this type of organization may have negative consequences on the interest of the task and on job satisfaction, but this is not the point here. The point is that it clearly shows the variability among the requests to be processed, and as a consequence, the variability of the reasoning processes involved. Operators deal with that variability by attempting first to apply directly the search procedures; requests are processed more fully only in case of failure of the initial attempts.

3.2.2   Simplifying by using a canonical model of the interlocutor

Another aspect of communication competence is described in a study of the verbal activity of the secretaries of a medical centre (Falzon et al., 1986). These secretaries are in charge (among other activities) of taking appointments for some of the specialists of the centre. The study focused on the activity of two of the secretaries, equal in domain knowledge, but differing in experience as operators.

An interesting result of this study concerns the dialogue strategies. The experienced operator's dialogue strategy enables her to reach very quickly (in usual cases) an agreement on a date and time of appointment. The expert strategy has two main characteristics:

— First, it is based on a very early hypothesis on the nature of the request: this hypothesis is made on the grounds of a model of the standard dialogue partner (i.e. the usual client). This model postulates that the client wants an appointment for himself/herself or for a close relative, that the appointment is wanted as soon as possible, and that the client is in a standard administrative situation as far as the Sécurité Sociale (Medical Welfare) is concerned.

— Second, it uses dialogue speed-up mechanisms, allowing some conversation steps to be avoided. For example: if an hypothesis of 'request of appointment with specialist X' has been made, do not ask for an explicit confirmation of the hypothesis, but propose, a date and time of appointment (on the basis of the above model). In general, it is more effective to act as if the interpretation is correct: if it happens to be wrong, it will provoke a negative and focused reaction of the user; if it is correct, it will speed up the dialogue.

The interest (for computer understanding) of using a model of the standard dialogue partner has been stressed by Shapiro and Kwasny (1975), Leroy (1985) and Lubonski (1985). The use of early interpretations of the queries in order to speed up the dialogue has been advocated by Hayes and Reddy (1983).

The strategy can be summarized as follows:

Assume the caller to be a 'standard client' (do not test directly this hypothesis). As soon as possible, propose a solution:

— if the solution is accepted, then the hypothesis is validated and the task is completed;

— if the solution is rejected, then apply recovery procedures.

Although the expert strategy is very efficient in most cases, it may lead the operator along wrong paths in case of very unusual requests. For instance, if the client does not want an immediate date of appointment the operator has to consider the date constraints of the user. The operator may experience some difficulty in some cases in getting rid of her erroneous interpretations. The recovery procedures she uses prove sometimes ineffective when the actual case is too far from the model.

3.2.3 Simplifying by modelling the interlocutor

The preceding paragraphs described a situation in which a canonical model of the interlocutor is used. More complex situations may occur, as will be shown in the study of an activity of diagnosis by telephone. The study has been conducted in a firm which constructs and sells programmable controllers. Some engineers of the firm are in charge of answering telephone queries of clients who experience difficulties in using the equipment they have bought. The equipment is of varying complexity, and the callers have various levels of competence in the domain. The task of the engineers is to diagnose the origin of the client's problem.

Table 1. The consequences of modelling the interlocutor's domain knowledge

	What is the level of competence in the domain of the information seeker ?	
Estimated competence	low	high
Consequences on dialogue management	Take control of the dialogue as soon as possible Use yes or no questions	Do not interrupt the information seeker; wait for a question
Consequences for diagnosis	The problem is (probably) not a complex one	The problem may be complex
	The problem is probably caused by one of the usual errors of the novice users	The problem is not a routine one
	The error will be found in the way the preliminary operating procedure has been completed, or in the user's program	The problem may be originated by a hardware failure, or by an error in the designer's program
	Check if the tests described in the operator's manual have been completed	The tests have probably already been done and do not provide any useful information

Telephone conversations have been recorded and transcribed. Interviews have been conducted with the engineers, focusing on the way they handled these conversations. A second interview was conducted, at least two months after the recordings, using the following method: the engineer had to read, line by line, the transcription of his conversations, and was asked to verbalize all he could say about the dialogue. The goal of this procedure was originally to gain some insight in the way the diagnosis was performed. The striking result is that many of the verbalizations do not concern the problem, but the level of competence of the client. These evaluative remarks occur very early, after having read a few lines of transcription. Evaluation seems to be made using two kinds of information:

— the content of the client's utterances: the technical concepts mentioned by the client, the correctness of the linking between concepts, etc;
— the way these concepts are expressed by the client: hesitations for example are noticed.

Evaluation is then made both on the grounds of technical knowledge and verbal behaviour. This evaluative activity has consequences also both on diagnosis and dialogue management, as Table 1 shows.

These observations show that the assumption of a complete independence of domain competence and communication competence is very much debatable. In this situation, communication competence relies on domain competence, since the evaluation of the interlocutor's domain knowledge is, in part, based on the expert's technical knowledge.

These observations also indicate how modelling the interlocutor's domain knowledge helps in choosing an appropriate level of processing. In case of a 'low domain knowledge' evaluation, the request will probably require only a minimum processing. On the contrary, a 'high domain knowledge' evaluation implies a non-routine problem, thus a more complex processing.

3.2.4 Levels of processing

These three examples show an organization of the processing strategies which differs sharply from the functioning of present interactive systems. It indicates a possible structure for the understanding units of these systems, using two modes.

A superficial, economic mode should be enough to process quickly the standard routine cases (i.e. the more frequent cases). By economic mode, we mean an analysis based on the use of knowledge structures strongly dependent on the domain, the task, and the prototypical stiuations. This mode could proceed by searching for keywords, sufficient to evoke schemata corresponding to the more usual problems, these schemata being then used to guide the comprehension process. Syntactic analysis, for example, can be simplified to a large extent.

It is only in some cases that the system would switch to a more elaborated processing mode. This mode would use:

— more complex programs of phonetic decoding (in the case of oral input), of syntactic and semantic analysis;
— other knowledge structures, less stereotyped and less dependent on the domain.

Of course, one can suppose that this second mode will be less economical not only in terms of the complexity of its subprograms, but also in terms of the necessary response time. But this is true also of human processing: rare problems take longer to solve than easy, usual ones.

In the situations presented above, the second mode is triggered in different ways:

— in case of a failure of the economic mode;
— in case of difficulties in applying the economic mode (signalled by a negative reaction of the interlocutor);
— in case of a diagnosis that the economic mode is not applicable (because of the nature of the interlocutor).

In all these situations the elaborated mode is used only when necessary. This solution has notable advantages, particularly in terms of the low workload implied for the machine when the economic mode is sufficient (i.e. in a large number of cases).

However, a second solution is possible, in which the elaborated mode would operate permanently, off-line, observing the utterances of the user and the answers of the economic mode. This could allow not only for its intervention in case of failure or difficulties of the economic mode, but also for a 'meta-dialogue' behaviour, thus avoiding some communication errors, caused for example by the use of erroneous schemata.

3.3   Hierarchical Cognitive Processing

One aspect of cognitive economy lies, as we have seen above, in the ability to use different modes of reasoning, as a function of the complexity of the problem. A second aspect of cognitive economy lies in the ability to choose an appropriate level to process the problem, in order to lower the cost of processing. The goal of this choice is to maximize the number of situations where operative or routine knowledge structures can be used.

An example of this second aspect is presented in a study of the functioning of an administrative information centre (Roman Parré, 1984). In this centre, which can be consulted by telephone, two categories of operators are involved, generalists and specialists. Calls are first processed by generalist operators, in charge of answering the more usual queries (i.e. to process the routine requests), and of transferring the calls to the adequate specialist operator. Generalists must then know the answers to the more usual queries, be able to detect questions outside their competence, and be able to choose the relevant specialist. This choice is made on the basis of some keywords found in the caller's initial request. Concerning this process, the author spots two interesting elements:

— Understanding, for the generalists, does not mean being able to give a definition of the keywords, but being able to categorize these words. Generalists have to be able to react to very different specialized vocabularies. In some cases, they know the meanings of these words; but it may also happen that, without really knowing the meaning of a word, they are still immediately able to specify the domain to which it belongs, and then transfer without hesitation the call to the appropriate specialist.
— The keywords for the generalist differ from the keywords for the specialist. This means that the elements used for categorizing the query are not those which allow the elaboration of an answer. In other words, the operative processing schemata of the specialist and of the generalist are not the same. This is of course because their tasks differ.

The task could have been organized in other ways. There could be a single category of operators, who would receive and process all calls, whatever the nature of the query. But, considering the extent of the domain, this would mean that, in a large number of cases, these operators would be in a problem-solving situation, facing totally new requests. Thanks to the organization described above, the operators are always able to use operative knowledge structures (except in very unusual cases), and then proceed quickly and efficiently (as far as the mental processes involved are concerned).

The operators (generalists or specialists) use the same (operative) mode of functioning, but with different knowledge structures. What we see here is not (as it was in the case above for telephone information operators) the allocation of different modes of reasoning to different operators, but rather the allocation of different hierarchical steps to different operators.

These results provide hints for the design of intelligent interactive systems. They suggest a multiexpert system structure, in which one of the experts would be in charge of allotting the job to the other experts of the system. Here again, this is quite different from what we usually observe in the functioning of present understanding systems. In most cases, these systems behave like an all-knowing operator, expert in several domains. For this reason, the specific competence consisting in determining the domain of expertise which must be triggered does not exist in these systems. As we have seen, understanding systems are not economical because they use a single mode of functioning. A second drawback is that they are not given the capability of operating at the level of abstraction at which economical procedures can be used.

REFERENCES

Amalberti, R., Carbonell, N., Falzon, P. and Jollivet, C. (1986). *La communication orale homme–homme: un modèle de référence pour la communication orale homme–machine?* Report to the ATP 'ARI Communication', December 1986.

Bailey, G. D. (1985). *The effects of restricted syntax on human–computer interaction* (Report MCCS-85-36). Las Cruces: New Mexico State University, Computing Research Laboratory.

Benbasat, I., Dexter, A. S. and Masulis, P. S. (1981). An experimental study of the human computer interface. *Communications of the ACM*, 24 (11), 752–62.

Bevan, N., Pogbee, P. and Somerville, S. (1981). MICKIE—A microcomputer for medical interviewing. *International Journal of Man–Machine Studies*, 14, 39–47.

Borenstein, N. S. (1986). Is English a natural language? In: *Foundation for Human–Computer Communication*. K. Hopper and I. A. Newman (eds). Amsterdam: North-Holland.

Card, W. I. and Lucas, R. W. (1981). Computer interrogation in medical practice. *International Journal of Man–Machine Studies*, 14, 49–57.

Chapanis, A. (1978). Interactive communication: a few research answers for a technological explosion. *Proceedings of the 86th Annual Convention of the American Psychological Association*, Toronto, Ontario, Canada.

Chin, D. (1984). An analysis of scripts generated in writing between users and computer consultants. *AFIPS Conference Proceedings*, 53, 637–42.

Clancey, W. H. (1983). The epistemology of a rule-based expert system. A framework for exploration. *Artificial Intelligence*, 20, 215–51.

Clark, H. H. and Wilkes-Gibbs, D. (1986). Referring as a collaborative process. *Cognition*, 22 (1), 1–39.

Evans C. R. (1976). Improving the communication between people and computer. In: *Proceedings of the NATO ASI on Man–Computer Interaction*, Alphen aan den Rijn, Netherlands.

Falzon, P. (1983). *Understanding a Technical Language* (Research Report 237). Rocquencourt: INRIA.

Falzon, P. (1984). The analysis and understanding of an operative language. In: *Proceedings of INTERACT '84, 1st IFIP Conference on Human–Computer Interaction*, 4–7 September 1984, London.

Falzon, P. (1989). *Ergonomie cognitive du dialogue*. Grenoble, France: Presses Universitaires de Grenoble.

Falzon, P. (in press) Cooperative dialogues. In: *Modelling Distributed Decision-making*. J. Rasmussen, J. Leplat and B. Brehmer (eds). Chichester, UK: Wiley.

Falzon, P., Amalberti, R. and Carbonell, N. (1986). Dialogue control strategies in oral communication. In: *Foundation for Human–Computer Communication*. K. Hopper and I. A. Newman (eds). Amsterdam: North-Holland.

Gould, J. D., Lewis, C. and Becker, C. A. (1976). *Writing and Following Procedural, Descriptive, and Restricted Syntax Language Instructions* (Report RC 5943). Yorktown Heights, New York: IBM Watson Research Center.

Hauptmann, A. G. and Green, B. F. (1983). A comparison of command, menu-selection and natural-language computer programs. *Behaviour and Information Technology*, 2 (2), 163–78.

Hayes, P. J. and Reddy, D. R. (1983). Steps towards graceful interaction in spoken and written man–machine communication. *International Journal of Man–Machine Studies*,. 19, 231–84.

Hendler, J. A. and Michaelis, P. R. (1983). The effects of limited grammar on interactive natural language. In: *Proceedings of CHI '83—Human Factors in Computing System*, A. Janda (ed.), 12–15 December 1983, Boston, MA, USA.

Krauss, R. M. and Glucksberg, S. (1977). Social and nonsocial speech. *Scientific American*, 236 (2), 100–5.

Krauss, R. M. and Weinheimer, S. (1964). Changes in reference phrases as a function of frequency of usage in social interaction: a preliminary study. *Psychonomic Science*, 1, 113–14.

Kupka, I. and Wisling, N. (1980). *Conversational Languages*. New York: Wiley.

Leroy, C. (1985). Structure des requètes. In: *Analyse linguistique d'un corpus d'oral finalisé*. Report to the GRECO Communication Parlée (GRECO 39). M.-A. Morel et al. (eds). Paris: IRAP.

Lubonski, P. (1985). Natural language interface for a Polish railway expert system. In: *Natural Language Understanding and Logic Programming*. V. Dahl and P. Saint-Dizier (eds). Amsterdam: North-Holland.

Meunier, A. and Morel, M.-A. (1987). Les marqueurs de la demande d'information dans un corpus de dialogue homme–machine. *Cahiers de Linguistique Française*, 8.

Morel, M.-A. (1985). Analyse linguistique d'un corpus d'oral finalisé (Centre de renseignements SNCF à Paris). In: *Analyse linguistique d'un corpus d'oral finalisé*. Report to the GRECO Communication Parlée (GRECO 39). M.-A. Morel et al. (eds). Paris: IRAP.

Ogden, W. C. and Boyle, J. M. (1982). Evaluating human–computer dialog styles: command vs form/fill-in for report modification. In: *Proceedings 26th Annual Meeting of the Human Factors Society*. 25–29 October 1982, Seattle, Washington, USA.

Onorato, L. A. and Schvaneveldt, R. W. (1986). Programmer/nonprogrammer differences in specifying procedures to people and computers. In: *Empirical Studies of Programmers*. E. Soloway and S. Iyengar (eds). Norwood, NJ, USA: Ablex.

Potosnak, K. M. (1984). Choice of interface modes by empirical groupings of computer users. In: *Proceedings of Interact '84*, 4–7 September 1984, London.

Ramsey, H. R. and Grimes, J. D. (1983). Human factors in interactive computer dialog. In: *Annual Review of Information Science and Technology*, Vol. 18. M. E. Williams (ed.). White Plains, NJ, USA: Knowledge Industry.

Richards, M. A. and Underwood, K. (1984). Talking to machines: How are people naturally inclined to speak? In: *Contemporary Ergonomics 1984*. E. D. Megaw (ed.). London: Taylor & Francis.

Roman Parré, M. (1984). La fonction de renseigner. In: *La communication dans la ville: du discours municipal à la mission de renseigner*. Paris: INA.

Scapin, D. L. (1986). Intuitive representations and interaction languages: an exploratory experiment. In: *MACINTER I*. F. Klix (ed.). Amsterdam: North-Holland.

Shapiro, S. C. and Kwasny, S. C. (1975). Interactive consulting via natural language. *Communications of the ACM*, 18 (8), 459–62.

Spérandio, J. C. and Létang-Figeac, C. (1986). *Simulation expérimentale de la synthèse vocale en dialogues oraux de communication homme–machine. Etude ergonomique*. Report to the GRECO Communication Parlée (GRECO 39). M.-A. Morel et al. (eds). Paris: IRAP.

Zoltan-Ford, E. (1984). Reducing variability in natural language interactions with computers. In: *Proceedings 28th Annual Meeting, of the Human Factors Society*, Vol. 1. Santa Monica, CA, USA.

SECTION 2
LEARNING PROCESSES

HUMAN LEARNING OF HUMAN-COMPUTER INTERACTION: AN INTRODUCTION

Y. WAERN

Department of Psychology, University of Stockholm,
S106-91 Stockholm, Sweden

1 HISTORIC BACKGROUND

The topic of learning is probably the single most researched topic in psychology. Is there anything new to add to that topic where human–computer interaction is concerned? Can't we just apply the knowledge already gained to this new field?

Some experiences from prior research in learning indicates that concepts and principles derived in one particular field cannot simply be transferred to a new field. One of the most extensive programs in the study of learning was introduced by Ebbinghaus (1885). The general idea was to construct a quantitative theory of learning, with the nonsense syllable as the smallest common unit. Amount of material learnt was related to learning time in different conditions. Later it was found that learning meaningful material could not be described by the same concepts as those used in describing learning of nonsense syllables. The prior knowledge of the learner, structured in schemata, tended to assimilate the material studied to those schemata (cf. Bartlett, 1932).

Another extensive research tradition was concerned with the effect of motivation and reinforcement on learning. More or less involved theories aimed at relating the response strength during learning to these factors. Although it can never be denied that motivation and reinforcement are important factors in learning, the pure quantitative approach is too meagre to explain the complexity of intellectual learning. There was a big leap in terms of number and type of concepts required when research turned to studying concept learning (cf. Bruner et al., 1956). Their qualitative approach showed that learners formed hypotheses and tested these during learning. Again new concepts were needed to describe learning in a problem-solving context (Anzai and Simon, 1979). Learners were found to evaluate not only the result of the problem-solving as such but also the process of deriving the result.

COGNITIVE ERGONOMICS:
UNDERSTANDING, LEARNING AND DESIGNING
HUMAN–COMPUTER INTERACTION

2 STRATEGIES AND CONCEPTS IN STUDYING LEARNING IN HUMAN–COMPUTER INTERACTION

The topic of human–computer interaction can thus be expected to require new concepts, new learning principles. At the same time we can expect at least some old principles to be valid. Two different research strategies can be used: one consists in studying the applicability of 'old' concepts and principles to the new field, the other consists in studying the new field as it is, redetect the 'old' principles as well as detect new principles. Both strategies have their strengths and weaknesses. The strategy of taking 'old' concepts as point of departure has the advantage of directly linking new facts to old knowledge, but the disadvantage of risking missing new facts which require new concepts. The strategy of taking the situation as such as point of departure has the advantage of handling things at their proper level and complexity, but the disadvantage of not knowing how the situation chosen can be linked to other situations (the generality problem) and thus also not knowing how to relate the knowledge gained to already existing knowledge and to other developing knowledge.

In the research of learning within the field of human–computer interaction we find both strategies. In the chapters presented in this Section we shall mainly meet the first strategy.

Learning computerized tasks has some aspects in common with procedural learning, as is pointed out by Allwood in the chapter which follows. A novice user starts learning a new system by applying knowledge outside of the system. The application of prior knowledge to the system to be learnt can be regarded as a problem-solving process. First the problem has to be understood, then learning can proceed by forming associations between specific goals and successful procedures (specific commands). Commands serving different goals have to be distinguished from each other. Later, the specific procedures are combined to procedural chunks, or methods, by which a sequence of commands are combined to achieve a specific goal. At the expert stage, the plans are combined to achieve large-scale goals, and the conditions for selecting the most appropriate method in the particular situation for achieving that goal are specified.

We thus see that there are two aspects of learning which have to be considered in a human–computer interaction situation. The first concerns what the learner already knows. The second concerns what the learner has to learn. If the learner in some way can utilize prior knowledge in the new situation, the learning can be facilitated to the degree that the prior knowledge complies with the requirements of the new situation. This insight has been used in several attempts at analysing the learning situation in human–computer interaction as well as in suggesting adequate instruction for this situation. Relating what should be learnt to prior knowledge is an old and well-founded educational recommendation. At the same time, no new knowledge would ever develop if we were totally dependent on our old knowledge. It can then be asked how prior knowledge can be modified and how new knowledge can be added to the old. In Table 1 some different ways of using prior knowledge are listed, together with the requirements found regarding the facilitation of learning. These aspects are relevant to the first phases of learning. Later, in Table 2, some different ways of modifying knowledge during later phases of learning will be presented. It should be noted that since computer systems are so complex, the different learning phases may be intermingled for different portions of the system.

Table 1. Different uses of prior knowledge and their requirements for facilitation of learning

Type	Requirement for facilitation of learning
Transfer	Easily discernible similarity between old and new situations Common elements
Metaphor	Familiarity Relevance Validity
Inference	Relevance Validity
Metaknowledge	Validity

Research relevant to the different uses of prior knowledge and the different modifications of knowledge will be presented, and the chapters which follow in this Section will be related to these tables.

2.1 Use of Prior Knowledge

Let me first present an overview of different types of use of prior knowledge suggested to be relevant in human–computer interaction.

2.1.1 Transfer

The idea of transfer rests on some simple assumptions. First it is assumed that some relationship exists between old knowledge and the material to be learnt. Second, it is assumed that the old knowledge which is related to the new situation is transferred to the new situation. The strength of transfer is dependent on the similarity between the old and the new situation as well as the strength of the knowledge to be transferred (cf. Waern, 1985). Transfer is thus related to a quantitative conception of learning. However, some qualitative aspects are included through the assumption of the similarity relationship.

The assumptions allow the researcher to predict the learning performance if he/she can specify the knowledge that is needed to perform a particular task as well as the knowledge related to the new task that learners have already acquired.

The transfer approach to learning is one example of how concepts from previous learning research have been successfully applied to the field of human–computer interaction. In particular, the definition of 'similarity' as 'number of common elements' has been applied in several different studies (Polson and Kieras, 1985; Polson et al., 1986; Polson et al., 1987; Vossen et al., 1987; Ziegler et al., 1986; Ziegler, Vossen and Hoppe, this volume). By objective analyses of the task to be learnt, as well as of tasks previously learnt, it has been found that the rate of learning is more or less linearly related to the number of new elements (here rules for how actions are related to goals) to be learnt. This finding supports the idea that the elements that already have been learnt are 'transferred' to the new situation without any further effort.

When the old task is totally contained in the new task, this finding is probably valid. However, earlier research also indicates that there are some common elements that will interfere with learning rather than facilitate learning. When there already exists knowledge which conflicts with the material to be learnt, new learning will be hampered by the need to repress or unlearn old knowledge. In particular, old knowledge elements which contain the same conditions as the new situation but actions other than those required in the new situation will cause interference.

Interference effects have been demonstrated within the field of human–computer interaction (cf. Waern, 1985; Rosson and Grischkowsky, 1987). It is not easy to overcome these effects. Even learners who are aware of the interference cannot at the beginning counterbalance the effect of their interfering knowledge. This finding indicates that learners are more dependent on their prior knowledge, at the beginning at least, than on their strategies for learning.

This means that to predict the learning of a particular computer system interference in a real world situation we should be able to tell what prior knowledge will be used by different users in a particular situation. This prediction is complicated by the individual differences to be found, not only in knowledge but also in the approaches to the task. Control of individual approaches for predictive purposes may be impossible outside the laboratory. However, some post hoc explanations support the contention of the importance of individual prior knowledge.

Let us look at the studies presented in this Section by Ackermann, Stelovsky and Greutmann in the perspective of transfer. These researchers suggest that the interaction between human and computer may differ with the task, the dialogue, and the individual. In the transfer analyses performed by Kieras and Polson and others following their tradition, the task and the dialogue are confounded in the objective analysis of the system to be learnt. However, a particular user may have knowledge about the task but not about the dialogue, or about the dialogue but not the task. Different transfer effects should then be expected.

Ackermann et al. studied the effects of dialogue grammars which were unfamiliar to their subjects. They found great individual differences. In one case the difference was traced back, not to knowledge about the system, but to knowledge about the task. The task consisted in drawing a car with a particular drawing system. A subject who performed this task better than anybody else was found to have drawn cars since he was a child. This result shows that in a transfer analysis it is important to consider task knowledge independent of system knowledge.

In another case the same researchers found that the difficulty of using a pocket calculator could be explained by the dialogue grammar needed to handle it. The calculator used a so-called 'polish notation', which is very different from and conflicting with most people's notions of pocket calculators.

However, Ackermann et al. also found that knowledge alone could not be responsible for some of the effects found. So for instance, in a rather unfamiliar situation, where a robot should be instructed to perform a particular task, great individual differences were again found. Some users extended the given grammar, others restricted it. Since both the task and the dialogue grammar were unfamiliar, it would be difficult to predict the differences from different prior knowledge. Instead they suggest that the variation could be due to different cognitive styles, motivation, memory capacity and other individual skills.

The conclusion from the transfer analysis is that in designing a system which is easy to learn the designer has to consider prospective users' previous knowledge of the dialogue grammar as well as of the task. There is no dialogue grammar which is optimal for all tasks and for all users.

2.1.2 Metaphors

In the transfer studies, similarity (often in terms of common elements) between the old and the new situation is essential. However, in the new, complex field of human–computer interaction, application of prior knowledge probably proceeds in other ways than through direct transfer. A fuzzier and vaguer way to use prior knowledge is to use analogical thinking. Analogies to familiar situations, however far-fetched, can be used to guide the users' expectations of and learning in the new situation. Analogies and metaphors in learning a computer system have thus represented important new concepts to be used in understanding both the ease and the difficulty people have in learning new computer tasks (cf. Carroll and Thomas, 1980; Halasz and Moran, 1982).

I shall now consider the use of metaphors, and thereby include the concept of analogy. The use of a metaphor implies a mapping between two domains: the source domain (the metaphor itself) and the target domain (the domain to which the metaphor is applied). There is no particular requirement that the two domains should be similar. The creation and understanding of a metaphor can require far-fetched inferences, and is not dependent on any simple triggering of prior knowledge (cf. Black, 1962).

The use of metaphors in human thinking seems to be very common, and there is reason to believe that people learning computer systems spontaneously create metaphors to describe their conceptions of the systems, and that their learning may be influenced by the metaphors that are suggested to them. Metaphors can, as well as transfer from prior knowledge, be either helpful or hindersome. Metaphors that are relevant and valid for the task to be performed as well as familiar to the user are helpful. Metaphors which suggest to users concepts and ideas in conflict with the system to be learnt are hindersome.

In this Section, two chapters are particularly concerned with metaphors and analogies. One, by van der Veer et al., addresses the question of how metaphors can be designed for helping people to learn computer systems. The other, by Allwood, deals with the question of identifying analogies which can be regarded as having caused errors during users' learning of computer systems.

Let me start with the educational aspect of metaphors. There do not exist any general rules of how metaphors should be chosen and expressed to be helpful. Instead, the construction of metaphors is more an art than a science. However, van der Veer analyses some aspects of this art to gain more insight into it. Since a computer system is a complex concept, it is probable that no single metaphor can cover all aspects of it. Thus, van der Veer has constructed metaphors, covering different levels of the system, from a general, task-related level, down to the specific keystrokes needed for a particular procedure. The metaphors can be conveyed both in talking about the communication—i.e. by explicit metacommunication—and in actual communication—i.e. in the words or icons chosen.

Let us next turn to the possibility that analogical thinking introduces errors in the learning of a system. This topic is dealt with by Allwood, who is very careful about the diagnosis of errors as 'analogical'. Since analogical reasoning does not require any particular similarity between the source and the target domain, any error could, if we wished, be regarded as analogical. Allwood uses a stricter criterion, requiring a rather low level of abstraction for the analogy. Also, errors clearly caused by other factors are not regarded as analogical. Allwood identifies different sources for the analogies: outside the system (a typewriter analogy) and inside the system (analogies with other programs or within a program in the system). Interestingly enough, he finds that most errors during learning are due to analogies within the system itself. This suggests that a user of a computer program will restrict use of prior knowledge to the knowledge he/she finds most relevant. Analogies from sources outside the system are clearly only relevant at the very beginning of learning, when no other sources exist. As soon as the learner gets to know something about the system, the system itself will serve as a more important source for analogies.

To conclude, metaphors and analogies represent important approaches to the learning of computer systems. In designing systems to be easy to use it is important to find a relevant, valid and familiar metaphor to start with. For further learning of the system, it is as important to design systems which are consistent, so that learners may capitalize upon analogical thinking also within the system.

2.1.3 Inference

Whereas transfer and metaphors refer to a mapping between two static domains, i.e. the domain of 'old' and 'new' knowledge, next use of prior knowledge concerns the transformations of prior knowledge which are possible through the use of inferences. Inferences can serve both to derive new conclusions from old knowledge elements and to interpret experience gained during learning.

The inferences users draw by working with a system are often hasty and based on too little evidence. This is probably due to the character of the learning. There is often no relevant feedback, and the user often forgets exactly what actions he/she has performed. Instead, the user constructs more or less well-founded hypotheses, which can be supported or rejected by further experience with the system. It is therefore to be expected that the inferences can be compatible as well as incompatible with the system during learning. The probability that the user will make inferences which are compatible with the system should increase during learning.

The incompatible inferences can be seen as a new source of errors. How can such incompatible inferences be studied? One example is seen in the chapter by Wilson, Barnard and Maclean in this Section. These researchers asked people to explain depicted states of the system to a friend. It was found that some explanations were correct, that others were incorrect and repeated, but also that new incorrect explanations were given in conflict with other facts which were earlier explained in a correct way. Wilson et al. suggest that the underlying knowledge from which the correct explanations were derived may remain compartmentalized and functionally inaccessible in a novel context. This finding indicates that users' learning does not consist in a simple addition of facts to a knowledge base. Another suggestion is that the explanations given by the subjects reflect an hypothesis testing, similar to that found in concept learning experiments. Correct hypotheses may then not always be detected, or some may be forgotten once voiced.

The topic concerning inferences has hitherto been much less researched than the earlier discussed topics of transfer and metaphors. It will be interesting to see if human–computer interaction researchers will take up the challenge to study this topic, and whether the result will be of any use to system designers. There are already some signs of a new development of concepts, represented by the work by Lewis et al. (1987).

2.1.4 Metaknowledge

An important factor in determining how people learn consists in their knowledge about their own knowledge, i.e. their metaknowledge. If they see at what points their knowledge is incomplete or incorrect, they will have greater chances of selecting the appropriate new knowledge to complete or correct the old knowledge.

It can then be relevant to ask whether or not people know what they know and what they are uncertain about. One way of studying this is to tell people to ask questions, when they are uncertain. This approach is illustrated by Allwood's study, where subjects were invited to ask questions while reading a manual. In this study, the total number of questions asked was not related either to performance on the computer tasks or to the subjects' own ratings of the difficulty of the manual. It was further found that different types of questions were related to performance in the tasks, however, in a way that is very difficult to interpret. It can be concluded from this study that people who only read a manual will not be able to predict the difficulties to be encountered during their attempts at performing a task in the system.

Therefore, they will not be able to ask questions which are relevant and which can help them in performing their task. Experience with the system is essential both for learning and for understanding where the difficulties may lie.

This conclusion is supported by the study presented by Schindler and Schuster in this Section. Here, different groups received different amounts of information in three different manuals. The subjects were encouraged to ask questions while they performed different tasks in the computer systems. It was found that for the very first task performed, the subjects asked fewer questions, the more information provided in the manual. Since the information in the most detailed manual was to some extent unnecessary, this finding is somewhat surprising. Maybe the subjects got overwhelmed by the amount of information and could not see anything more to ask about. Such an interpretation is compatible with Allwood's finding that subjects asked some irrelevant questions, related to hardware aspects.

In a later phase in Schindler and Schuster's study, in which the subjects were asked to perform tasks in the system by instructing the experimenter, no differences in the number of questions asked were found between the groups having different manuals. This result suggests that the impact of the information in the manual is greatest in the very first beginning of learning. Later, the experience with the system itself will pose problems which require other types of questions than those based only on the manual.

Further work relating metaknowledge both to experienced difficulty and to information given is necessary to tell when users spontaneously would seek help and what kind of help they would then find useful. Such a knowledge is needed to design help messages in answer to requests from the user (passive help systems) or when the user seems to need them (active help systems).

Table 2. Some different ways of modifying knowledge and factors facilitating the modification

Modification of knowledge	Facilitating factors
Generalization	Similarity
Discrimination	Easily discernible distinction and feedback
Restructuring	?
Addition	Repetition
Compilation	Repetition
Automatization	Repetition

2.2   Modification of Knowledge

Let us now turn to questions related to how knowledge is modified during learning. Table 2 presents some different ways in which knowledge may be modified, together with the conditions that have to be met to affect the modification.

2.2.1   Generalization and discrimination

These learning principles are derived from concept learning studies, and have also been suggested as important principles for procedural learning (Anderson, 1982). The generalization principle implies that procedures or rules related to specific knowledge elements may be generalized to other, similar knowledge elements. This principle is balanced by the discrimination principle, where procedures or rules related to different knowledge elements are distinguished. Through different outcomes the learner will find that the same general knowledge cannot be applied, and thus be forced to make a discrimination between the knowledge elements.

The generalization principle is similar to both the transfer and the metaphor principles discussed above. The discrimination principle requires both that the domains are clearly recognized as different and that the feedback is unequivocal. These principles have not been explicitly focused by researchers in human–computer interaction, perhaps because they can be incorporated under the transfer and metaphor principles.

One paper in the present volume can, however, to some extent be related to the discrimination principle, i.e. the study presented by Schindler and Schuster. In this study, one group was only instructed in the procedures to be used to perform different tasks, whereas two groups were given labels for these procedures in addition. The labels give, according to Schindler and Schuster, a more precise explanation of the dialogue steps to be performed. This is exactly what is meant by creating a clear distinction between different knowledge elements. In the study it was found that the groups instructed by labels also tended to use these labels to a greater extent, at least in the first stage of learning. Later no differences were found, which also could be expected. Once learners have learnt that knowledge elements should be distinguished from each other, several bases for the distinction can probably be found.

2.2.2   Restructuring

More radical modifications of knowledge are required if not only a hierarchical relationship is considered as in generalization and discrimination, but also other kinds of relationships. The psychological researchers who have been mostly concerned with the questions of the structure of knowledge are the Gestalt psychologists. The insight achieved by these early researchers should be of great value in the field of human–computer interaction. Few researchers in this new field have attended to the principles found by the Gestalt psychologists. The general ideas of creating compatibility between system and user models to facilitate learning and use of computers can, however, to some extent be seen as intimately related to the structure principle in Gestalt psychology.

The structure principle implies that a problem can find its solution only by a structural approach. A piecemeal approach cannot lead anywhere, if the pieces cannot be placed in a relevant structure. Several problems are difficult to solve, only because the formulation of them induces us to build a structure which is not compatible with the problem solution (Wertheimer, 1959).

It can be acknowledged that it is difficult to see the parallels between such classical problems as that of the area of parallelograms or the Galileo problem and the problems involved in human–computer interaction. Most problems in the latter field may really be solvable by a piecemeal approach. However, programming seems to be a field where the structural approach may be more relevant.

Let us therefore consider the study presented by Samurçay in this Section, in the light of a structural approach. In the study some novice programmers were given the task to learn PASCAL programming and produce different programs, giving the arithmetic sum of some numbers. The solution of the problem requires a loop-plan, a good representative of a particular structure. According to Samurçay, the construction of the loop-plan involves the identification of what shall be repeated and the condition for when the repetition should end. It is suggested that the different loops which can be constructed in PASCAL will pose different difficulties because of their different structures.

In some cases, the novice programmers are required to restructure the thinking about the problem. Although Samurçay does not call her analysis structural, the concepts studied and the observations performed can be regarded as different ways of defining the structure. By this analysis she is able to predict what aspects of the programming problem will cause novices most difficulties and what aspects will be easy to understand. The data support her hypotheses.

It might be fruitful to reintroduce the concept of structure and restructuring into the field of human–computer interaction. Hereby it would also be necessary to specify some concepts more than was necessary for the problems hitherto approached. For instance, the very perceptual approach taken by the early Gestalt psychologists should be supplemented by a more conceptual frame of reference. There are certainly conceptual 'Gestalts' as well as perceptual ones!

2.2.3 Additions to knowledge

Much of the earlier work in learning psychology has been concerned with how new knowledge is added to old. Not only nonsense syllables have been studied but also other kinds of verbal learning (cf. Anderson and Bower, 1973).

In the field of human–computer interaction, pure verbal learning may only concern the learning of meaningless connections between command names and their corresponding functions, a situation which nowadays gets less and less common. Since the situation in human–computer interaction is so much more complex than in the verbal learning laboratory, it is probable that the apparently meaningless relations can be meaningfully interpreted, either through transfer or metaphors. Since there are no chapters dealing with this topic, I shall not dwell longer on the concept of 'pure' additions to knowledge.

2.2.4 Compilation and automatization

One conspicuous aspect of procedural learning consists in the decrease in time taken to perform different tasks. The change in performance time follows the 'power law' of practice, suggested to be applicable not only to motor skills, where it was originally suggested, but also to cognitive tasks such as justifications for geometric proofs (Neves and Anderson, 1981).

It has been proposed that the decrease in performance time is due to three main learning principles, corresponding to different stages of learning; the first related to understanding the task and eliminating errors, the second related to the compilation of procedural chunks, and the third related to automatization of the procedures in each chunk (cf. Anderson, 1982).

The chapter by Wilson, Barnard and MacLean addresses the question of how performance time is related to different aspects of the tasks learnt. Since the learning session in their study was rather short, mainly the first learning principle, i.e. elimination of errors, can be supposed to be applicable. The results show that the performance time indeed decreased. However, the decrease in performance time was only valid for three tasks, whereas the performance time for six tasks remained constant. As expected, the decrease in learning time was mainly due to elimination of errors. However, all errors were not eliminated. Instead the decrease was related to the number of so-called 'major reattempts', i.e. actions concerned with handling errors by restarting the process of solving the problem from the beginning. During learning, more local error corrections were performed. From this result it can be suggested that learning proceeds with different rates for different aspects of what has to be learnt, for instance different command sequences. An important finding is that the elimination of errors proceeds through different stages.

It can be concluded that the learning of a computer system is so complex that the issues of compilation and automatization will be relevant only for the most simple tasks (for instance data entry), and that the learning of a computer system in its entirety will be much more concerned with understanding the system and eliminating errors than with compilation and automatization of procedures.

2.3 Methodological Considerations

Most researchers know that the observation methods used influence the insights to be gained from their investigations. This is of course also true for research in human–computer interaction. The methods in turn are based on more or less explicit assumptions about the nature of the phenomenon to be studied. We found that transfer studies had to define the similarity between old and new situations very carefully, whereas metaphor studies had to rely on vaguer notions. A researcher who wants to attribute to the learners the capacity of drawing common sense inferences has to struggle with even fuzzier evidence.

In general, it seems that studies of learning in the context of human–computer interaction are based on fuzzy concepts and qualitative data to a much greater extent than studies of learning in other contexts. It can then be asked if the variability in results which are often difficult to interpret is due to the methods used rather than to the nature of the problem. The chapter in this Section by Wilson et al. addresses this question. By using different methods for assessing the knowledge of their subjects, these researchers found some converging evidence for their conclusions about learning. They found that performance in a real task and answers to a verbal questionnaire were highly related. The most conspicuous exception concerned the measure of the speed of keying in a data entry task. This result induces us to reflect on how different types of measures relate to each other and to underlying theoretical concepts.

At present, researchers seem to have different opinions concerning what should be focused in the research on human–computer interaction. Some propose to approach the problem on a low, 'keystroke'-level. These are represented by Card et al. (1983). Others propose a more general approach, concerned with 'understanding' and 'pleasure'. These are represented by the researchers around Donald Norman (cf. Norman and Draper, 1986). Both approaches are probably necessary in order to understand the full range of phenomena in learning computerized tasks.

3 CONCLUSION

It was proposed at the beginning of this introductory discussion that the new field of human–computer interaction would be expected to require new concepts and new approaches where learning was concerned. Has this proposition been supported by the analyses performed and the studies presented?

It can be concluded that some earlier principles indeed are valid in this field too. In particular, the transfer principle has found support in several studies. However, it can also be concluded that the complexity of the situation cannot be adequately accounted for by applications of simple, quantitative principles. Number of common elements may predict ease of learning, but number of new elements will not be enough to predict learning time. Conflicting knowledge can hamper learning to an unpredictable extent, and prior knowledge of a more far-related kind can be used both to facilitate and hinder learning. Some of the new concepts required are concerned with metaphors and analogies as well as weak inference rules.

When modifications of prior knowledge are concerned, the simple quantitative power law of practice is certainly too crude. A more qualitative approach is required to capture the phenomena of generalization, discrimination and restructuring. It will for instance probably be necessary to develop theories of attention which are more suited to deal with complex, knowledge-dependent situations than the presently available ones. Discriminations and restructurings require appropriate attentional focus, a concept which hitherto has been hardly touched on in human–computer interaction research.

Since much of the complexity can be reduced by finding an appropriate structure, it might also be possible to develop new, conceptual, Gestalt laws which are suitable for this particular field. From a research point of view, the reduction to the learning of keystrokes or command sequences may be feasible but not very fruitful.

When systems design is concerned, the present practice which is compatible with our knowledge about human learning is mainly restricted to the use of metaphors. We have seen that metaphors represent an important shortcut to learning at the first stage of learning. However, it must be carefully considered what content should be mediated by the metaphor and in what form the metaphor should be presented. Also, since systems often are learnt piecemeal, metaphors should encourage the combination of pieces. Analogies within a system will after the first stage of learning be more important than analogies between the system and the world outside the system.

Systems design has hitherto to a very small extent considered the support for modification of knowledge, i.e. the learning of new aspects, new procedures. Since computer systems after all are designed to offer something new, it seems odd to support nothing but the relation to a familiar world without computers. People are able to learn new things, both by generalization and discrimination, by restructuring and addition. They certainly need support in these situations. This can be provided by immediate and adequate feedback, by directing their attention to the right place, by explanations and by repetitions. The challenge to stimulate detection of new features, understanding of new structures and learning of efficient procedures is there—the questions of what will work and how still await answers.

REFERENCES

Anderson, J. R. (1982). Acquisition of cognitive skill. *Psychological Review*, 89, 369–406.

Anderson, J. R. and Bower, G. H. (1973). *Human Associative Memory*. Washington: Winston.

Anzai, Y. and Simon, H. (1979). The theory of learning by doing. *Psychological Review*, 86(2), 124–40.

Bartlett, F. C. (1932). *Remembering*. Cambridge: Cambridge University Press.

Black, M. (1962). *Models and Metaphors*. Ithaca, New York: Cornell University Press.

Bruner, J. S., Goodnow, J. and Austin, G. A. (1956). *A Study of Thinking*. New York: Wiley.

Card, S. K., Moran, T. P. and Newell, A. (1983). *The Psychology of Human–Computer Interaction*. Hillsdale, NJ, USA: Erlbaum.

Carroll, J. M., and Thomas, J. C. (1980). *Metaphor and the Cognitive Representation of Computing Systems*. (Report RC 8302). New York: IBM Watson Research Center.

Ebbinghaus, H. (1885). *Ueber das Gedächtnis*. Leipzig: Duncker.

Halasz, F. and Moran, T. P. (1982). Analogy considered harmful. *Proceedings of Human Factors in Computing Systems*, Gaithersburg.

Lewis, C., Casner, S., Schoenberg, V. and Blake, M. (1987). Analysis-based learning in human–computer interaction. In: *Human–Computer Interaction—INTERACT '87*. H.-J. Bullinger and B. Shackel (eds). Amsterdam: North-Holland, pp. 275–80.

Neves, D. M. and Anderson, J. R. (1981). Knowledge compilation: Mechanisms for the automatisation of cognitive skills. In: *Cognitive Skills and Their Acquisition*. J. R. Anderson (ed.). Hillsdale, NJ, USA: Erlbaum.

Norman, D. A. and Draper, S. (eds) (1986). *User Centered System Design*. Hillsdale, NJ, USA: Erlbaum.

Polson, P. G. and Kieras, D. E. (1985). A quantitative model of the learning and performance of text editing knowledge. *Proceedings of CHI '85*, pp. 207–12.

Polson, P., Muncher, E. and Engelbeck, G. (1986). A test of a common elements theory of transfer. In: *Human Factors in Computing Systems. Proceedings of CHI '86*. M. Mantei and P. Orbeton (eds). New York: ACM.

Polson, P. G., Bovair, S. and Kieras, D. (1987). Transfer between text editors. In: *Human Factors in Computing Systems and Graphics Interface. Proceedings of CHI + GI '87*, J. M. Carroll and P. P. Tanner (eds). New York: ACM, pp. 27–32.

Rosson, M. B. and Grischkowsky, N. L. (1987). Transfer of learning in the real world. In: *Human–Computer Interaction—INTERACT '87*. H.-J. Bullinger and B. Shackel (eds). Amsterdam: North-Holland, pp. 891–6.

Vossen, P. H., Sitter, S., and Ziegler, J. E. (1987). An empirical validation of cognitive complexity theory. In: *Human–Computer Interaction—INTERACT '87*. H.-J. Bullinger and B. Shackel (eds). Amsterdam: North-Holland, pp. 71–5.

Waern, Y. (1985). Learning computerized tasks as related to prior task knowledge. *International Journal of Man–Machine Studies*, 22, 441–455.

Wertheimer, M. (1959). *Productive Thinking*. New York: Harper & Row.

Ziegler, J. E., Hoppe, H. U. and Fähnrich, K. P. (1986). Learning and transfer for text and graphics editing with a direct manipulation interface. In: *Human Factors in Computing Systems. Proceedings of CHI '86*. M. Mantei and P. Orbeton (eds). New York.

LEARNING AND USING TEXT-EDITORS AND OTHER APPLICATION PROGRAMS

Carl Martin ALLWOOD

Department of Psychology, University of Göteborg, Box 14158, S400–20 Göteborg, Sweden

1 INTRODUCTION

Interacting with a computer for the first time has been compared to getting to know a foreign culture (Galambos et al., 1983). Getting to know a new culture means learning many seemingly arbitrary conventions. A characteristic difficulty for the newcomer in this connection is to know where the similarities between the known and the new culture end and where the dissimilarities begin. Another characteristic difficulty for the novice in a new culture is realizing which things are central in the new culture and which are peripheral. Thus, the novice will often not know when it is important to find out more about a phenomenon and when it is not so important.

The present chapter discusses a series of studies by our research group[1] which investigate the above-mentioned aspects of novices learning a new application program. First, as a general frame of reference, novices' knowledge, its development and use will be considered.

2 NOVICES' KNOWLEDGE, ITS DEVELOPMENT AND USE

An individual's knowledge is usually seen as consisting of at least three different knowledge types: skills, conceptions and episodical memories. Often, the last two types of knowledge are called declarative knowledge.

The development of skill knowledge has been well described in a more general context by Anderson (1982). According to Anderson, the novice, when carrying out new tasks uses declarative knowledge to a large extent. To carry out the task, the novice constantly has to retrieve declarative knowledge from long-term memory. Experts, on the same occasions, use highly

---

[1]Carl Martin Allwood, Mikael Eliasson and Torbjörn Wikström

COGNITIVE ERGONOMICS:
UNDERSTANDING, LEARNING AND DESIGNING
HUMAN–COMPUTER INTERACTION

integrated procedural chunks, not demanding much interaction with long-term memory once the relevant procedural chunk has been activated.

Kay and Black (1985) summarized a series of empirical studies on the development of skill in text editing and arrived at a description consistent with that of Anderson. Initially, the novice can only access computer commands through knowledge of natural language. At this stage, the novice will choose commands with names resembling the names of functions from everyday life (for example typewriter operations) which have effects similar to those that the user wants to carry out. Initial learning consists of the novice building up associations between specific goals and specific commands, or between a specific goal and a number of commands, each leading to that goal. Next, the user builds up integrated chunks which incorporate the knowledge necessary to carry out simple plans. Such procedural chunks consist of a specific goal and a sequence of commands that will achieve that goal. Finally, at the expert stage, the plans are combined to achieve large-scale goals, and application conditions are integrated into the plans so that in a particular context a suitable plan can be chosen. It seems likely that the same development also takes place in connection with the learning of other types of application programs such as database programs and spreadsheet programs.

The users' conceptual knowledge is important in a number of ways. Previous research has shown that an adequate conceptual model of the system can help a person to generalize his/her knowledge and thus enable the person to handle new, and hitherto unencountered situations. Furthermore, the user may be more able to realize when an error has been made and, perhaps more importantly, a conceptual model can help the user to find ways to recover from error situations.

The development of users' conceptual understanding of the system has not been studied to the same extent as has the development of skill knowledge. A number of studies have shown that the novices' conceptual knowledge of the system is fragmented (see Allwood, 1986). This is understandable when considering how development in novices' conceptual models of the system can come about. First, the user may draw inferences during a particular experience of the system. Lewis (1986a,b) has discussed how users with no prior domain-specific knowledge construct conceptual models of the system by drawing inferences when watching a concrete interaction with the system. For this purpose, Lewis argued, novices only have very general and weak rules of inference at their disposal. Since research results by Lewis and his co-workers (i.e. Carroll and Mack, 1984) and others, have shown that novices draw hasty and often erroneous conclusions from too little evidence, conceptualizations developed in this way, i.e. as part of novices' first conceptualizations of how the system works, will often be inadequate.

Second, the user may gradually develop a conceptual model of how the system works by abstracting similarities and noticing dissimilarities between episodic memories of different interaction experiences of the same system (see Kolodner, 1983). Since many experiences of approximately the same type will be needed before the user can make suitable generalizations and discriminations, it may take a long time before the user, in this way, arrives at an adequate conceptual model of the system. Third, the user may receive information from another person, i.e. a teacher, as to what is an adequate conceptual model of the system. Research results in cognitive psychology suggest that it is unrealistic to expect the user to take over directly the model of the system as the instructor intends it to be comprehended. The user understands in terms of his/her own prior conceptions and the instructor is often unable to predict what conceptions the user will use when comprehending the information given, Thus, this way of developing a model of the system will also often, at least initially, lead to inadequate conceptualizations.

Other studies have attempted to characterize the novices' way of interacting with the computer when using application programs. One general conclusion is that the novices' way of interacting can best be described as problem-solving (Card et al., 1980; Waern, 1984). This is in line with Anderson's description of the development of skill knowledge. As noted above, previous research has shown that novices, for example, when text editing, tend to draw too hasty conclusions based on too little evidence (see also, Hammond and Barnard, 1984). Thus, on top of having a poor conceptual model to start from, novices' computer interaction is also often characterized by seemingly poor information processing strategies. However, as pointed out by Carroll and Mack (1984), it may be that the novices' strategies are actually adaptive if one considers the learning situations novices often find themselves in.

Such learning situations are usually characterized by the novice being given too much information to handle at the same time (see also Carroll and Carrithers, 1984). In such a situation, making mistakes and errors both with respect to the commands issued and the content added to the conceptual model is perhaps the only or the easiest way of making any progress at all. The alternative may be to become bogged down in thought. There is obviously a delicate balance between making many costly errors and not making so many errors but at the same time getting less experience with the system.

What actions the novice will perform when interacting with the program are determined by the specific knowledge that is currently activated in the novices' memory. The content currently activated in memory depends on a number of factors, for example the content of the novices' stored knowledge, how this content is stored in memory (for example, what knowledge is stored functionally close to what other knowledge) and the external situation that the novice is in. From this perspective it is natural to expect that the novice's errors will often be generated by multiple simultaneously acting causes.

3 THE EFFECT OF ANALOGICAL THINKING ON NOVICES' USE OF APPLICATION PROGRAMS

In the two first of the three studies in the series to be reviewed (Allwood and Wikström, 1986; Allwood and Eliasson, 1987), we have studied the role of analogical thinking in the generation of novices' errors when they interact with the computer. First some general points about analogical reasoning and its study.

Analogical thinking can be defined as thinking that involves the application of knowledge about one domain, i.e. the source domain, to another domain, the target domain. Since there is no necessary minimal semantic distance that has to hold between the source and the target domain it is not surprising that many authors have characterized the essence of thinking as analogical (i.e. Lakoff and Johnson, 1980; Carbonell, 1981). Furthermore, although it is clear that the analogical nature of thought is to our advantage much, perhaps most of the time, in our studies we have focused on the negative effects of analogical thinking.

It is not always clear when an error should be classified as due to misplaced analogical thinking. For example, before classifying an error as due to analogy, we need to be reasonably sure that the knowledge analysed as inappropriately applied was indeed part of the novice's knowledge store when the error was made. Even disregarding this problem, it is not always clear whether a specific error is to be regarded as originating from a specific source domain or not. Owing to these problems it may not be possible to list the set of all reasonable, but mistaken, analogies from a specific source domain to a target domain, and only these. However, such an attempt has been made by Douglas and Moran (1983) with respect to the source domain of typewriting and a specific text-editor, EMACS, in the target domain of text editing. These authors used two criteria of similarity when they generated the set of operator analogies: (1) the operators utilize the same key in both domains, and (2) the operators have similar effects in both domains.

The danger that the researcher might miss reasonable, but mistaken, analogies is presumably greater than him/her including unreasonable ones. However, one case of possible over-inclusion in Douglas and Moran's list will be discussed here to illustrate the difficulties involved. The example concerns an error where the subject intends to insert a character into a word and places the cursor in the wrong position. This error was classified as due to a misplaced typewriter analogy by Douglas and Moran. However, this action would have led to an error also in a typewriting situation since, when typewriting, the text is added directly at the current place of the typehead. Accordingly, it is probably best not to see this error as an example of the typewriter analogy. A possible counter argument is that with modern typewriters one has to position the writing head one step to the right of the point where the text is to be added and then press a special key which activates a white correction-tape so as to reverse back to the position where the correction is to be made. However, this counter argument may not be valid because if this interpretation was correct, the subject, finding him/herself one step too far to the right of where the correction is to be made, would back the cursor to the correct position instead of directly attempting to add the character to the text, which is what actually happens in these situations. This example also indicates that in the future it may be important to take the specific typewriters that subjects have interacted with into account, due to the development of different versions of modern typewriters.

In other cases the problem may be that the suggested analogy is vague. For example, it is not clear if using 'Return' (only) when loading the operator system should be seen as an analogy to pressing the carriage return in order to insert the paper into the typewriter. In general, it appears to be a good methodological strategy to allow only analogies which are relatively concrete, since at an abstract level almost anything can be analogous to something else.

Yet another difficulty is that source domains may overlap and that accordingly it may be difficult to identify an error uniquely with a specific source domain. For example, both the typewriter and computer program may use a specific convention not valid in the particular program being used. When a person erroneously uses the convention, it is not clear what source domain is involved, the typewriter or the first computer program, or both. In the present studies this problem has been dealt with by identifying the difficulty with both, or all, of the relevant source domains focused on in the study.

The point with this discussion is not to argue against attempts in research on analogical thinking to list possible analogies between source and target domains. The point is that we should be very careful before we see such lists as final, or methods to generate such lists as algorithmic. A better way to regard such attempts is as most likely incomplete and perhaps in need of alteration. Furthermore, it should be noted that the analyses performed in connection with the effect of analogical thinking in the studies reviewed here and in the study by Douglas and Moran (1983) are not strong enough to prove in any conclusive sense that the analogies identified have caused the errors analysed. However, it seems very likely that the analogies identified have contributed to the errors and difficulties analysed.

In their study, Douglas and Moran (1983) analysed the role of the typewriter analogy in generating the errors that four novices made in early interactions with the EMACS text-editing program. The authors compared the novices' errors to a list of errors generated by using the method described above. The result showed that about 60 per cent of the novices' errors appeared to have occurred due to misplaced use of the typewriter analogy. To understand this result better, it should be noted that about 30 per cent of the novices' errors were classified as syntactic in nature and that none of these errors was classified as due to the typewriter analogy. Thus, 83 per cent of the observed semantic errors were analysed as due to the typewriter analogy. Furthermore, inefficient commands (i.e. commands that bring the subject closer to the goal but are not the most efficient way of reaching the goal) were not analysed.

In our first study to be reviewed (Allwood and Wikström, 1986), we analysed the difficulties experienced by four subjects as they attempted to learn two application programs; Multiplan, a spreadsheet program and Condor, a relational database management program. Two subjects learned one of the programs and two subjects the other, aided only by instruction manuals. The instruction manuals contained detailed instructions on how to perform different exercises. Towards the end of the manuals, the instructions became somewhat less detailed. We asked the subjects to 'think aloud' as they read the manuals and carried out the exercises on the computer. In addition, the contents on the computer screen were recorded. None of the subjects had any experience with the type of program they were learning but all of them had several years' experience with text editing and statistical analysis programs.

In the study, we analysed the effect of three source domains on subjects' difficulties. These were: (1) other parts of the currently used program; (2) other computer programs that the subject was acquainted with, and (3) knowledge about typewriters and the paper and pencil domain.

Three general criteria were used when identifying misplaced analogical thinking. According to the first of these criteria, the analogy has to be the main cause of the subject's difficulty, otherwise it was not judged as due to analogical thinking. The second criterion is only relevant in connection with analogies from the currently used program. This criterion says that if the subject believes that he/she is in a part of the program where it would have been correct to issue the erroneous command, then the erroneous command is not considered as due to analogical thinking. In these cases the error is seen as due to the subject's mistaken belief about his/her location in the program. The third criterion rules that the analogy should be on a relatively low level of abstraction.

A total of 489 episodes containing subjects' difficulties were analysed. Of these, 17 contained inefficient commands. The analyses showed that subjects' difficulties were of many different kinds. The subjects also showed great variation with respect to the number of difficulties experienced. For example, one of the two subjects learning the spreadsheet program had 338 episodes containing difficulties, whereas the other subject learning the same program had only 92 episodes. It was possible to relate the difference between the subjects with respect to the number of difficulties experienced to differences in learning style, i.e. differences with respect to how closely the subject attempted to follow the instructions in the manual.

With respect to misplaced analogical thinking, the analyses indicated that only 23 per cent of the 489 episodes analysed were related to any of the three sources of analogy considered. Analogies from the program currently used were responsible for 15 per cent of subjects' difficulties. Knowledge concerning other computer programs was analysed to be responsible for 8 per cent of subjects' difficulties and knowledge of the typewriter and the paper/pencil domain for 6 per cent of subjects' difficulties. These figures do not add up to 23 per cent since subjects' difficulties were occasionally analysed as deriving from more than one source domain.

Owing to the difficulties in getting an overview of the subjects' knowledge of natural language, we did not systematically analyse the effect of knowledge from this domain on subjects' difficulties. However, a few obvious examples of such difficulties will be given. In written natural language 'x/y' often means 'either x or y'. A difficulty relating to this convention is that subjects when they saw 'C/R' in the database program (meaning 'carriage return') thought they had to select either C or R. Likewise, in the spreadsheet program, subjects interpreted 'Alpha/Value' as meaning that they had to indicate their choice of either Alpha or Value. In fact, the message means that if the user presses a numerical key (and one of a few other keys) the Value mood will be entered, otherwise the Alpha mood will be entered. Accordingly, no explicit choice is needed. As a last example, in the spreadsheet program, of two alternatives, the alternative put in parentheses was the default alternative. However, the subjects, in accordance with conventions in written natural language, interpreted the alternative not in parentheses as the default alternative.

Conclusions

Our study suggests that analogical thinking played a smaller role for our subjects' difficulties compared with the results presented by Douglas and Moran (1983). However, it should be noted that the results in these two studies are not strictly comparable since our definition of difficulty encompassed more phenomena than did the definition of error used by Douglas and Moran. Two further differences between our study and that of Douglas and Moran are probably more important in explaining the differences in the results observed. First, our subjects had more general computer experience than Douglas and Moran's subjects and, accordingly, our subjects may not have been as prone to analogical errors as the subjects in that study. Second, our subjects learned programs where the typewriter analogy may have been less tempting than was the case for the text editing program studied by Douglas and Moran.

The second study to be reviewed (Allwood and Eliasson, 1987) analysed novices' errors when, for the first time, they attempted to carry out four elementary text-editing tasks in the WordStar program. A total of 28 subjects participated in this study, 8 of whom were secretaries. The subjects had had very little computer experience and 12 of them had never come into contact with a computer before. On average the subjects had only used a computer on one or two previous occasions.

The data in the study described were collected as part of the third study to be reported below. Here, only the parts of the procedure that are relevant to the present study are mentioned. The subjects first filled out a questionnaire on background variables. Secondly, subjects were asked to read a short instruction manual which first gave general information about microcomputers and word processing. Next the manual contained specific information about how to use limited parts of the operating system (DOS) and of the WordStar text editing program. In total, the subjects learned about 26 commands. The subjects were asked to read the instruction manual in such a way that they would later be able to answer some questions about the text and to perform some text editing tasks on the computer.

In the next step of the procedure of relevance here, the subjects were asked to carry out a computer interaction task. This task involved 17 subtasks. These included inserting the disks into the disk drives, loading the operating system, loading the WordStar program and the text file to be edited, carrying out four prescribed corrections on the text file, saving the edited file, exiting the WordStar program and the operating system and, finally, taking out the disks from the disk drives. The four editing tasks involved inserting missing characters into words, breaking up one paragraph into two and replacing three sentences with a new sentence. (The procedure and the statistical analysis of the results are described in detail in Allwood and Eliasson, 1987.)

The subjects were instructed to 'think aloud' and to report all keys pressed on the computer keyboard as they carried out the computer interaction task.

If a subject failed to report a key pressed, the experimenter did so instead.

The analysis involved all commands analysed as erroneous or inefficient. A command was classified as an *error* if it did not to any extent bring the subject closer to the goal-state for the task at hand. A command was classified as *inefficient* if it was not erroneous but was not on the shortest solution path to the goal. An example of an inefficient command is moving the cursor characterwise when wordwise is more efficient. In some cases it turned out to be problematic to establish the most efficient way to carry out the four text editing tasks, for these we only analysed subjects' deletion of characters from the text.

In total, 245 errors and 83 inefficient commands were found in the data. The subjects' erroneous commands were analysed into 104 *syntactic* errors (when the system does not respond at all or gives an error message) and 141 *semantic* errors (when the command is syntactically correct but does not bring the subject closer to the goal-state). Twenty of the subjects' syntactic errors and 30 of their semantic errors were separated from these classes of errors to form a new category called *physical execution* errors. These errors were analysed as being due to subjects' inability to physically master the keyboard. Examples of physical execution errors were errors where a subject did not press two keys simultaneously when this was required, and errors due to subjects' inability to handle the repetitive function of the keyboard.

Analogy analysis

In this study, two source domains of analogies were investigated, the program currently used and typewriter knowledge. Owing to subjects' lack of knowledge of other computer programs, this source of analogies was not analysed. Erroneous analogies from the program currently used were identified in two ways. When the subject had correctly used a command (or part of a command) and then used it erroneously, a priming analogy error was identified. Second, when a subject used a command before the appropriate situation was at hand, an anticipation analogy error was identified.

In some cases it was considered to be of interest to analyse not just the entire character string in the command but also the two components which together make up a command in the WordStar program. These two parts are: (1) the command group selector, a part that is common for a group of commands, and (2) the command indicator, a specific, goal-determined part that indicates a specific command in the group of commands identified by the command group selector.

All possible analyses were not reasonable to perform in relation to all error categories and all ways of identifying analogies. The restrictions imposed on the analogy analysis are marked with '–' in Table 1. In the priming analogy analysis for syntactic errors, only those components of the command string that were erroneous in relation to the subject's goal were analysed. For semantic errors in the priming analogy analysis, the entire command and the command indicator were analysed. However, the command group selector was not analysed since this part of the command presumably does not carry any clear meaning for the subjects. Of the physical execution errors, only those which would have been syntactically correct had they been correctly executed were included in the priming analogy analysis, viz the 12 cases where the subjects pressed the control key and the alpha-character in succession instead of simultaneously. If the erroneous execution method had previously led to correct results, the error was identified as a priming analogy. It did not make sense to perform the priming analogy analysis with respect to the other 38 errors in this category. Inefficient commands were analysed as due to priming analogy only if the entire command string had previously been correctly used.

In the anticipation analogy analysis, only subtasks involving actions leading to the loading of the WordStar program and subtasks involved in saving the text file and exiting were included. The subtasks involved in performing the actual text editing tasks were not included in this analysis since these subtasks are not likely to occur as either source or target domains for anticipation analogies. One reason for this is that the goals for these subtasks are very distinct and very different from the goal in the preceding and the following substeps. Furthermore, the instruction manual did not detail how to carry out the subtasks involved in the text editing tasks, and subjects are thus not likely to have had a clear representation of the commands necessary for carrying out these tasks. In contrast, the manual gave a very detailed description of the steps necessary to carry out the other parts of the computer interaction task. Further restrictions on the anticipation analogy analysis are that syntactic and semantic errors were not analysed with respect to the command group selector parts due to the high probability of a chance correspondence with the same part in a command to be used later on in the interaction task. For obvious reasons, the physical execution errors and the inefficient commands were not coded in the anticipation analogy analysis.

In the typewriter analogy analysis, syntactic and semantic errors were coded as due to this analogy if they were classified as typewriter analogies by Douglas and Moran (1983). This rule had three exceptions. First we excluded the error discussed above, involving insertion of characters into words, since this error is presumably best not regarded as due to the typewriter analogy. Second, we included errors where the subject substituted the shift key for the control key. The argument is that these keys are next to one another on the keyboard and that these two keys are not normally distinguished on typewriters. Third, we included errors where the subjects did not distinguish between the visual appearance or the meaning of the

Table 1. Results of analogy analysis for syntactic, semantic and physical execution errors and for subjects' inefficient commands (Ci = command indicator, Gs = group selector, – = no analysis performed)

	Syntactic errors	Semantic errors	Physical execution errors	Inefficient commands	Total
All errors or inefficient commands	84	111	50	83	328
Priming analogies					
Ci	2	0	–	–	2
Gs	25	–	–	–	25
Entire command	10	39	12	33	94
Anticipation analogies					
Ci	10	5	–	–	15
Entire command	14	0	–	–	14
Typewriter analogies	12	23	37	75	147
Percentage of errors determined	68%	37%	98%	92%	68%

symbol and the function of the symbol. An example of this error type is when subjects, when writing the file name substituted '1' (one) with 'l' (the letter l). Another instance of this type of error is when the subjects confused the control key with the key with the control symbol (⌃, issued by shifting a numerical key). The analogy behind errors of this type may have been: characters with similar appearance or meaning are interchangeable when typewriting and can be interchanged when text editing as well.

All physical execution errors that occurred because of holding a key down for too long were included in the typewriter analogy analysis. All inefficient commands were analysed with respect to the typewriter analogy. An inefficient command was coded as due to the typewriter analogy when there is no operator on the typewriter which is more efficient than the operator corresponding to the command used by the subject on the computer. A typewriter analogy was also identified when the subject solved a subtask in a more roundabout way than necessary, corresponding to how it would normally be performed on a typewriter.

Table 1 shows the result of the analogy analyses performed. It can be inferred from this table that 60 per cent of the subjects' errors were associated with at least one of the two main sources of analogy considered. When subjects' inefficient commands are included, this figure increases to 68 per cent.

In more detail, Table 1 shows that 68 per cent of the syntactic errors were associated with at least one of the three types of analogies investigated. Of the syntactic errors, 14 per cent, i.e. 12 errors, were associated with the typewriter analogy. This result does not actually contradict the finding of Douglas and Moran (1983) that no syntactic errors were associated with the typewriter analogy, since the only syntactic errors that were associated with the typewriter analogy in our data were errors where the subjects used the shift key or the key with the ⌃ symbol instead of the control key, i.e. errors which were not defined as due to the typewriter analogy by Douglas and Moran.

With respect to the semantic errors, 37 per cent of these errors were associated with at least one of the three analogies analysed. The typewriter analogy was associated with 21 per cent of the semantic errors (74 per cent of these were also associated with priming analogies).

Of the physical execution errors, 98 per cent were associated with either priming analogies or the typewriter analogy. Finally, 92 per cent of the inefficient commands were associated with either priming analogies or typewriter analogies. All the inefficient commands, except one, that were associated with one of the types of analogies were associated with both types of analogies.

It can be inferred from Table 1 that 36 per cent of the errors were associated with priming analogies, 12 per cent with anticipation analogies and 29 per cent with typewriter analogies. Furthermore, if the first two types of analogies were combined, the results show that 44 per cent of the errors were associated with analogies from the currently used program. When subjects' inefficient commands are included, 43 per cent of all erroneous or inefficient commands were associated with analogies from the currently used program and 45 per cent with the typewriter analogy.

To learn more about the nature of the analogies generated from subjects' knowledge about the program currently used, we analysed the number of subtasks occurring between the source of the priming and the anticipation analogies and the occurrence of the errors. For the errors analysed as deriving from priming analogies, the source of the analogy was within three subtasks from the occurrence of the error for 65 per cent of the errors. (Including the inefficient commands, the corresponding figure is 74 per cent.) For all errors associated with anticipation analogies, the source of the analogy was within two subtasks from the error.

In this study we also analysed the effect of some background variables on subjects' proneness to errors deriving from analogical thinking. The background variables investigated were: typing speed as self-estimated by the subjects, computer interaction experience, sex, age, grade in last school attended and level of education. A stepwise multiple regression analysis showed that typing speed was the variable that explained most of the variance with respect to the number of errors associated with misplaced analogical thinking ($R^2 = 0.38$, $p < 0.001$). Computer interaction experience was entered second into the analysis ($R^2 = 0.16$, $p < 0.05$). The more computer experience subjects had had, the less was their tendency to make errors due to misplaced analogical thinking.

Conclusions

Compared with the first of our studies reviewed (Allwood and Wikström, 1986), the conditions in the presently described study were more like those in the study by Douglas and Moran since a text editing program was used and the subjects had very little prior experience with computers. The results show that, with this type of program and subjects, misplaced analogical thinking plays a rather large role in the generation of subjects' erroneous and inefficient commands. Of these types of commands, 68 per cent were analysed as deriving from misplaced analogical thinking. The results also show that analogies from the currently used program appeared to play a somewhat larger role in the generation of subjects' erroneous commands compared with the typewriter analogy. When subjects' inefficient commands are included in the analysis, the role of analogies from the currently used program approximately equals that of analogies from the typewriter domain.

When comparing our results to those of Douglas and Moran (1983), it should be noted that we analysed more types of errors than Douglas and Moran did. When only subjects' syntactic and semantic errors are considered, our results indicate that only 18 per cent of these errors were derived from the typewriter analogy. It is not clear what factors caused the differences in results between our study and that of Douglas and Moran. However, the subjects in the Douglas and Moran study were all secretaries and all had had some training sessions on the computer before the study. Furthermore, Douglas and Moran used a greater range of editing tasks than we did. Finally, Douglas and Moran only studied four subjects whereas we studied 28 subjects.

It seems reasonable to assume that novices make more errors deriving from misplaced analogical thinking compared with more expert users. This is supported by the fact that even with the small range of computer experience represented among the subjects in our study, degree of computer experience was significantly related to frequency of errors derived from analogical thinking. Accordingly, we would expect the subjects in the study by Douglas and Moran to make less errors due to faulty analogical thinking since these subjects had somewhat more computer experience compared with the subjects in our study. However, the fact that all Douglas and Moran's subjects were secretaries, compared with 25 per cent of ours, suggests that these subjects should have more analogical errors than our subjects. The reason is that our results show that novices with high self-estimated typing speed were more prone to analogical errors, including errors due to typewriter analogies, than other subjects. Since typing speed explained more of the variance for analogical errors than computer experience did, it stands to reason that the subjects in Douglas and Moran's study made more analogical errors compared with our study. However, other differences between the two studies have probably also contributed since the discrepancy in the results in the two studies was quite large.

There are at least two routes to making it easier for subjects to learn application programs. One way is to improve the design of the programs to be learned. For example, the results discussed above elucidate the importance of consistency in program design, both within specific programs and between different application programs. However, the task of constructing programs that are easy to learn is difficult. For example, our results, when compared with those of Douglas and Moran (1983), indicate that differences in prior knowledge between different categories of novices are important in determining what types of errors novices make when text editing.

Another way to improve the situation for beginners is to improve the learning situation. Not much systematic research has been devoted to discovering what learning conditions lead to improved learning of application programs (but see Carroll and Carrithers, 1984; Carroll and Kay, 1985). Accordingly, it is important that this type of research is carried out to complement research on program features.

4 QUESTION ASKING AS A MEANS OF LEARNING TO TEXT-EDIT

Novices learn to use application programs in a variety of situations, formal and informal. In most of these, an important means for novices to improve their learning is to ask questions. For example, by asking questions novices can relate the information presented by the teacher to the understanding of the system they are attempting to construct. When trying to use the system, novices ask for advice from teachers, colleagues and at computer advisory centres. Question asking may also be of potential importance in the future when help facilities in application programs will be more intelligent and interactive than is the case today. In the last study to be reviewed (Allwood and Eliasson, 1988) we attempted to learn more about question asking as a possible means of learning application programs.

Only a few studies have investigated question asking in the context of learning about computers (Miyake and Norman, 1979; Coombs and Alty, 1980; O'Malley et al., 1984; Pollack, 1985). Miyake and Norman varied the difficulty of an instruction manual that novices read in order to learn a text editing program. The main result was the novices asked fewer questions on the hard version of the manual compared with the easy version. The other three studies (Coombs and Alty, 1980; O'Malley et al., 1984; Pollack, 1985) indicate that novices experience difficulties when they pose questions about programs to persons who know more than they do about computers. Coombs and Alty found that when asking advice at a computer advisory centre, novices had difficulties making themselves understood, due occasionally to their questions being too constrained and concrete. O'Malley et al. studied the interaction between one beginner and a teacher. They found that the beginner's questions were often irrelevant in relation to the learning goals of the teacher. The beginner's questions often concerned different details on the screen. Pollack found that an adviser who received written questions from users on an electronic mail system often provided answers that reinterpreted and went beyond the questions asked by the users.

In the present study, we investigated whether the opportunity to ask questions (to which the subject received answers) when reading an instruction manual would improve subjects' subsequent performance on the computer and on a test probing subjects' understanding of central concepts in the manual text. We further investigated the relation between how difficult subjects rated the manual text and the number and types of questions asked.

As noted above, the data in the present study came from the same experiment as the data used in Allwood and Eliasson (1987). Accordingly, only details that are relevant to the description of the present study will be added to the description of the experiment given above. The subjects were asked to read an instruction manual containing information about computers, word processing and the WordStar program in such a way that they would later be able to perform some text editing tasks on the computer and to answer some questions about the text. The instruction manual was divided into seventeen sections. After reading each section, the subjects rated the difficulty of the section on a five-point scale with the extremes 'very easy to understand' and 'very difficult to understand'.

After having rated how difficult the section was to understand, half of the subjects were instructed to ask questions on the section just read, whereas the remaining subjects were not given this opportunity. For each section, the subjects were given approximately one-third of the reading time for constructing questions. The same subjects were also encouraged, while reading the text, to write down any questions that they might think of. The experimenter attempted to answer the subjects' questions in a simple and helpful manner. A short dialogue sometimes took place after the experimenter's first answer since the experimenter attempted to see to it that the subject had understood his answer. Subjects not in the question asking group were compensated for the extra time used by the subjects asking questions by being given extra reading time.

After reading the instruction manual, the subjects carried out the computer text editing task and filled out a questionnaire designed to tap subjects' understanding of central computer concepts. Finally, the subjects were asked to fill out Eysenck's extroversion–introversion scale. The computer interaction task has been described above.

The questionnaire contained three sections. The first tested subjects' understanding of 11 single concepts, for example, screen picture, operating system, program, control key, cursor and file. In the second section, subjects were asked to describe the flow of information between the keyboard, RAM memory, disk and screen. The third section asked subjects to draw a hierarchical tree showing the relations between nineteen different computer concepts. Details on the scoring of the questionnaire data are given in Allwood and Eliasson (1988).

The main result of the study was that asking questions did not have any significant effect on subjects' number of errors or inefficient commands on the computer interaction task, nor on subjects' performance on the questionnaire testing text understanding. Finally, question asking did not significantly affect how subjects rated the text material in terms of difficulty. Thus, question asking did not make the manual easier to read according to subjects' ratings.

An obvious condition to be able to profit by asking questions, is to be able to formulate questions. The subjects showed great variation with respect to the number of questions that they asked. The total number of questions asked by the 14 subjects in the question asking group was 364. The total number of questions asked by individual subjects varied between 6 and 43. A correlation analysis showed that the only background variable (including subjects' results on Eysenck's extroversion–introversion test) that correlated with the number of questions asked was education level. Subjects with higher education asked more questions than subjects with lower education. However, there was no significant correlation between asking many questions and performing well on the two performance tasks. This result suggests that subjects, at least part of the time, asked suboptimal questions.

To follow up this result, we performed three kinds of analysis with respect to the questions asked. The subjects' questions were coded with respect to (1) content, (2) closeness to text in the manual, and (3) whether they contained any false presuppositions or not. The interjudge reliabilities for these three codings were 84 per cent, or more.

The type of questions most frequently asked concerned use of the WordStar program (168 questions). The other types of questions were ranked in order of frequency as follows: questions concerning hardware without specific relevance to text editing (75), questions concerning software with specific relevance to text editing (39), questions concerning the natural task domain, for example questions on office or orthographical terms (27), questions concerning software without specific relevance to text editing (23), questions without relevance to the task domain (writing and editing of texts with or without computers, use of computers in general) (20), and questions concerning hardware with specific relevance to text editing (11).

The results of the coding of the relation of questions to the manual text showed that 54 questions asked for information stated in the text, 64 questions asked for information inferentially close to the text and the majority of the questions (246) asked for information not judged to be inferentially close to the text.

The coding of whether the questions asked contained any false, explicitly mentioned assumptions, showed that the great majority (294) of subjects' questions did not contain any explicitly mentioned false assumptions, whereas 70 questions did.

Next, correlations were computed between number (and proportion) of questions of a specific kind and performance on the computer interaction task and on the questionnaire. The figures for subjects' number of errors in the computer interaction task were corrected to take account of the amount of help given to the subject by the experimenter and for the number of subtasks the subject had left unfinished or not attempted.

The results show that asking many questions about software aspects without specific relevance to text editing was related to having few erroneous commands in the computer interaction task ($r = -0.58$, $p < 0.03$). The correlation for number of questions asked concerning software aspects with specific relevance for text editing was in the same direction but was not significant. In contrast, a high proportion of questions on hardware aspects with no specific relevance to text editing was related to having many errors on the computer interaction task ($r = 0.55$, $p < 0.05$). No significant correlations occurred between degree of closeness of questions to content in text or number of questions containing false assumptions and number of errors in the computer interaction task.

The correlations between number (and proportion) of different question types asked and performance on the questionnaire followed the same pattern as that presented immediately above. However, in addition, the results showed that the more questions a subject asked concerning use of the WordStar program and the more questions a subject asked that were inferentially distant from the text, the better was the subject's performance on the task in the questionnaire where the subject had to describe the information flow between four components in the computer. Furthermore, the more questions the subject asked that contained false assumptions, the fewer items that subject drew into the hierarchical tree.

The correlations involving subjects' difficulty ratings of the text show that subjects who rated the text as difficult tended to ask more questions, the answers to which were stated in the text. No other question type was even close to being significantly correlated with the subjects' difficulty ratings. Furthermore, there was no significant correlation between the subjects' difficulty ratings and the number of questions asked. Finally, the results show that subjects who rated the text as more difficult tended significantly to have a poorer performance on both the computer interaction test and on the questionnaire, compared with subjects who rated the text as easier.

Conclusion

These results support earlier research, the general gist of which is that novices have difficulties asking questions which are helpful to their attempts to interact successfully with the computer. Our results suggest that the time subjects spent in this study asking questions and getting them answered might just as well have been spent on further encoding of the text.

It is of interest to ask why question asking did not appear to lead to improved learning. That there was no correlation between subjects' difficulty ratings and number of questions asked, suggests that subjects often did not ask questions when it would have been appropriate. The reason might be that they did not fully realize the need for clarifying information or alternatively that they realized the need but could not formulate the questions. In this context it is of interest that there was a positive correlation between rating the text as difficult and having poor results on the two performance measures. This suggests that subjects, to some extent, were able to detect when their understanding of the text was insufficient. However, this need not have indicated to the subjects the appropriateness of asking questions, perhaps because no topic in particular stood out as unclear and/ or because the subjects did not connect experiencing the text as difficult with it being a good strategy to ask questions. Further research is needed to evaluate these alternatives.

The results also showed that subjects who rated the text as more difficult than other subjects more often asked for information stated in the text than other subjects did. This suggests that the subjects who rated the text as difficult to understand, asked questions to a greater extent in order to improve their encoding of the text, whereas the function of question asking for subjects who rated the text as easier to understand tended to be to expand their understanding beyond the text.

The fact that there was no significant correlation between the number of questions asked and subjects' results on the two performance tasks suggests that even when subjects were able to formulate questions, this did not imply that they were able to ask about content that was relevant to improving their result on the two performance tasks.

The finding that subjects with poor performance results asked questions with different content compared with subjects with better performance results also merits attention. At least two explanations for this finding are possible. One is that subjects with poor computer performance had poorer knowledge of computers than the other subjects *and* that one cannot understand software aspects of the system until one has understood the hardware aspects. The second explanation focuses on subjects' strategical skills in text reading and question asking. According to this explanation subjects with poor computer performance asked the questions they did partly due to lack of skill in deciding what topics in the text are important to focus attention on and what questions are important to ask in relation to their goal. Although the instructions asked the subjects to read the text in such a way that they could perform text editing tasks on the computer and answer questions on the text, they may not have realized which type of questions were conducive to this goal. A combination of these two explanations might be that it was the poor performers' deficient knowledge about computers that made them think that it was a good strategy to learn about hardware aspects before software aspects.

The fact that there was no significant correlation between rating the text as difficult and asking the types of questions discussed, argues against explanations based on the premiss that poor performers asked the questions they did due to their having more deficient knowledge about computers than the other subjects. The skill explanation receives some support from the results that it was only level of education among the background variables that correlated significantly with the number of questions asked. One way to achieve the ability to formulate questions may be to take part in the activities in the classroom. This suggests that question asking, at least partly, is a matter of skill.

These results have implications for teaching and the design of computer programs. With respect to teaching, these results suggest that if question asking is to be an effective component in the beginners' learning process, many novices will need encouragement to ask questions and perhaps training in realizing when it is appropriate to ask questions. Furthermore, many novices may profit by explicit instructions concerning what topics are fruitful to ask about.

With respect to the design of computer programs, these results have important implications for the design of help facilities.[1] Current help facilities are usually not very helpful. For example, in the study by Allwood and Wikström (1986) reviewed earlier, we also studied the use that the two learners of the spreadsheet program made of the help facility in that program. Only one of the two beginners used the help facility at all on her own initiative. This user could only solve her problem after having read the help text on one of the eleven occasions she used the help facility (see also Houghton, 1984). The help facilities in the future will presumably be much more flexible than those of today. Carroll and McKendree (1986) and Houghton (1984) are amongst those who have contributed very interesting suggestions for how help facilities could be made more efficient.

The limitations in novices' question asking demonstrated in our study and in other studies is important in this connection. When clever design overcomes the current limitations in the computers' software for help facilities, limitations in the human 'software' may turn out to be the barrier for full use of the possibilities of getting help from the computer. Our study has shown that novices often may not realize when questions are appropriate and that even when they do, they may not be able to ask about the relevant matters. This suggests the limitation of passive help facilities and thereby suggests the potential importance of developing active help facilities for novice users.

---

[1] Thanks to Norbert Streitz for directing my attention to this relation.

5 CONCLUSIONS

The studies reviewed show the relevance of learning more about the different factors that influence novices' on-line attempts to understand and learn to use the computer. What factors determine what prior knowledge the novice will bring to bear in the interaction situation? When will novices' evaluative processes act to stop inappropriate use of prior knowledge? What factors determine which aspects of the application program and the computer system the novice will find relevant to understand further? It appears that it is questions such as these that we need to understand better if we want to give novices a more effective introduction to the computer and if we want to provide them with better programs to use. To really improve the situation for novices, it seems that we need to understand much better how novices use their mental resources, for example, their conceptions, skills and strategic knowledge, when they deal with the content of the application programs they are learning.

REFERENCES

Allwood, C. M. (1986). Novices on the computer: A review of the literature. *International Journal of Man–Machine Studies*, 25, 633–58.

Allwood, C. M. and Eliasson, M. (1987). Analogy and other sources of difficulties in novices' very first text editing. *International Journal of Man–Machine Studies*, 27, 1–22.

Allwood, C. M. and Eliasson, M. (1988). Question asking when learning a text-editing system. *International Journal of Man–Machine Studies*, 29, 63–79.

Allwood, C. M. and Wikström, T. (1986). Learning complex computer programs. *Behaviour and Information Technology*, 5, 217–25.

Anderson, J. R. (1982). Acquisition of cognitive skill. *Psychological Review*, 89, 369–406.

Carbonell, J. G. (1981). *Metaphor comprehension*. Department of Computer Sciences, Carnegie-Mellon University, CMU-CS-81-115.

Card, S. K., Moran, T. P. and Newell, A. (1980). Computer text-editing: an information-processing analysis of a routine cognitive skill. *Cognitive Psychology*, 12, 32–74.

Carroll, J. M. and Carrithers, C. (1984). Blocking learner error states in a training-wheels system. *Human Factors*, 26, 377–89.

Carroll, J. M. and Kay, D. S. (1985). Prompting, feedback and error correction in the design of a scenario machine. In: *Proceedings of CHI '85: Human Factors in Computing Systems*. L. Borman and B. Curtis (eds). San Francisco: ACM, pp. 149–53.

Carroll, J. M. and Mack, R. L. (1984). Learning to use a word processor: By doing, by thinking and by knowing. In: *Human Factors in Computer Systems*. J. C. Thomas and M. L. Schneider (eds). Northwood, NJ, USA: Ablex, pp. 13–51.

Carroll, J. M. and McKendree, J. (1986). *Interface Design Issues for Advice-Giving Expert Systems*. (Report RC 11984). New York: IBM Watson Research Center.

Coombs M. H. and Alty, J. L. (1980). Face-to-face guidance of university computer users—II. Characterizing advisory interactions. *International Journal of Man–Machine Studies*, 12, 405–29.

Douglas, S. A. and Moran, T. P. (1983). Learning text-editor semantics by analogy. In: *Proceedings CHI '83: Human Factors in Computing Systems*. A. Janda (ed.). New York: ACM, pp. 207–11.

Galambos, J. A., Wikler, E. S., Black, J. B. and Sebrechts, M. M. (1983). How you tell your computer what you mean; ostension interactive systems. In: *Proceedings CHI '83: Human Factors in Computing Systems*. A. Janda (ed.). New York: ACM, pp. 182–5.

Hammond, N. and Barnard, P. (1984). Dialogue design: Characteristics of user knowledge. In: *Fundamentals of Human–Computer Interaction*. A. Monk (ed.). New York: Academic Press, pp. 127–64.

Houghton, R. C. Jr (1984). Online help systems: A conspectus. *Communications of the ACM*, 27, 126–33.

Kay, D. S. and Black, J. B. (1985). The evolution of knowledge representations with increasing expertise in using systems. In: *Tutorial 11: Cognitive Issues in Interface Design, CHI '85*. J. B. Black (ed.). San Francisco: ACM.

Kolodner, J. L. (1983). Towards an understanding of the role of experience in the evolution from novice to expert. *International Journal of Man–Machine Studies*, 19, 497–518.

Lakoff, G. and Johnson, M. (1980). *Metaphors we live by*. Chicago: Chicago University Press.

Lewis, C. (1986a). A model of mental model construction. In: *Proceedings of CHI '86: Human Factors in Computing Systems*. M. Mantei and P. Orbeton (eds). Boston: ACM, pp. 306–13.

Lewis, C. (1986b). Understanding what's happening in system interactions. In: *User Centered System Design*. D. A. Norman and S. W. Draper (eds). Hillsdale, NJ, USA: Erlbaum, pp. 171–85.

Miyake, N. and Norman, D. A. (1979). To ask a question one must know enough to know what is not known. *Journal of Verbal Learning and Verbal Behaviour*, 18, 357–64.

O'Malley, C., Draper, S. and Riley, M. (1984). Constructive interaction: A method for studying user–computer–user interaction In: *Human–Computer Interaction—INTERACT '84*. B. Shackell (ed.). Amsterdam: Elsevier Science Publishers (North-Holland). pp. 1–5.

Pollack, M. E. (1985). Information sought and information provided: An empirical study of user/expert dialogues. In: *Proceedings of CHI '85: Human Factors in Computing Systems*. L. Borman and B. Curtis (eds). San Francisco: ACM, pp. 155–60.

Waern, Y. (1984). On the implications of users' prior knowledge for human–computer interaction. In: *Readings on Cognitive Ergonomics—Mind and Computers* (Proceedings of the 2nd European Conference). G. C. Van de Veer, M. J. Tauber, T. R. G. Green and P. Gregory (eds). Berlin: Springer-Verlag.

ACTION REGULATION AND THE MENTAL OPERATIONAL MAPPING PROCESS IN HUMAN–COMPUTER INTERACTION:

DOES THE INTERACTION REFLECT DIALOGUE GRAMMAR?

David ACKERMANN,* Jan STELOVSKY** and Thomas GREUTMANN*

*Work and Organizational Psychology Unit
ETH Zürich (Swiss Federal Institute of Technology),
CH-8092 Zürich, Switzerland
**Department of Computer Science, University of Honolulu,
Hawaii 94822, USA

1 THEORETICAL CONTEXT

1.1 Psychological Aspects

For an optimal development of the adult personality in working life it is indispensable that individual differences are taken into account as Ulich (1978, 1987) stated in his principles of differential and dynamic work design. There has to be the possibility to adapt and enlarge existing production systems sensu Ulich (or task-solving processes), to create new ones and switch between different ones thus permitting individual working styles. Triebe (1980) and Zülch and Starringer (1984) have shown that such individualized work can increase efficiency and productivity.

A theoretical explanation for the results mentioned is required. On the basis of the hypothesis of 'semantic information transfer' (Krause, 1982), Ackermann (1983) formulated the hypothesis that difficulties in human–computer interaction during task-solving processes are caused by the discrepancy between individual mental representations and cognitive styles on the one hand and the given operations of the scope of action prescribed by the software on the other hand. Therefore, cognitive and action regulation processes are the most important aspects in our research.

COGNITIVE ERGONOMICS:
UNDERSTANDING, LEARNING AND DESIGNING
HUMAN–COMPUTER INTERACTION

1.1.1 Cognitive processes

The 'mental representation' of tasks, tools and prescribed operation methods is involved in the ongoing task-solving process. External stimuli have to be transformed into an internal representation. Goals have to be derived and retransformed into actions. As a dynamic and task-specific process the mental operational mapping is defined by goals, scripts and plans (Schank and Abelson, 1977) and depends on motivation, knowledge in the sense of 'mental representation' and cognitive skills (Johnson-Laird, 1983). Oschanin (1976) called this mediating process 'operatives Abbildsystem' (task-related mental model) but did not formulate an explicit model. Ackermann and Stelovsky (1986) pointed out that the actual mental operational mapping process—as a cognitive process—is affected by individual differences.

1.1.2 Action regulation

Psychological action theory assumes that the basic unit of action can be described by a feedback loop (Miller et al., 1973). In the learning process, basic operations are combined into hierarchical organized action-schemes according to the learner's goals and semantic structuring of the task (Ackermann, 1987). In order to act, only the highest level of action-schemes (represented by chunks (Cube, 1968)) has to be consciously activated according to the goal of the action. We have investigated the composition of complex action-units and the resulting scope of action (Ackermann, 1986a,b, 1987). We could show that the mental operational mapping process and the resulting command programs are individually different and that problems in dialogue are due to mismatch between the mental operational mapping processes and the prescribed methods.

1.1.3 Model of human–computer interaction

We summarize the psychological theoretical background outlined above in the model of human–computer interaction depicted in Figure 1. During the task-solving process, data processing is the relevant common aspect of both the computer and the human. In the physical world, the computer offers tools to manipulate the physical data. The tools' user-interface can be seen as generated from a dialogue grammar formed by a set of rules for accepted input and generated output. This dialogue grammar defines the objective degrees of freedom in the physical scope of action. In this paper, we will concentrate on the input rules only. On the human side the dialogue grammar is perceived and matched with the task. This matching process is influenced by cognition, motivation, action knowledge and intentions. We assume that this forms a kind of syntax and semantics in generating actions and the observed actions can be described grammatically. The last step in the action regulation—the mental-operational mapping process—generates an observable result—the action sequence. Thus, the action sequence allows us to infer the subjective mental scope of action.

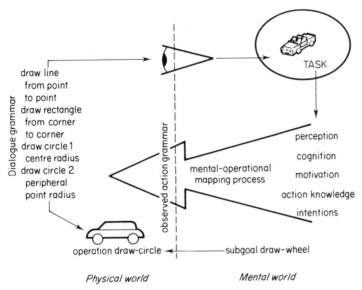

Figure 1. The model of human–computer interaction.

1.2 Computer Science Aspects

Human–computer interaction can be regarded as a communication process. Moran (1981) and Reisner (1981), for example, have created grammatical systems to describe the dialogue interface. For years, command languages have been the major dialogue scheme for the communication between man and the computer. Some research efforts have focused on the formal specification of a command language and its use as a system specification and design language. At present, representation of commands as menus and their selection by pointing (e.g. with a mouse) has captured the users' acceptance of interactive systems. To use icons and windows for representing data and command objects has come into fashion.

These approaches differ in how the objects are identified. In the command language approach, command names and parameters must be typed in to be identified. Because the command language is based on a grammar, it permits the specification of complex, structured commands. A complex grammar, however, imposes restrictions on possible user input. If the restrictions must be memorized, the cognitive complexity (Ackermann and Stelovsky, 1986) and thus also the mental workload increases. In a menu approach, commands labelled by names or icons are displayed on the screen and used for input by pointing. Since the dialogue grammar is simplified to a minimum, defining complex command sequences is tedious. Menus can be intuitive for the user, but by using them the user may underestimate the complexities of structure and grammar which are inherent in highly interactive applications. Moreover, a flat command structure that can restrict the subjective action-scope of the user (Spinas, 1986, 1987) came into fashion. In short, a command language is precise and expressive, but effortful to use if its grammar must be remembered.

Adaptable interactive systems allow us to offer varying dialogue grammars and investigate their influence on the observed action sequences, their structure and suitability for the task-solving process. Such experiments help us to develop and refine the requirements for the design of computer dialogues.

1.3 Goals of Our Research

Psychological aspects

As a consequence of the principles of differential and dynamic work design, computer dialogues must be adaptable according to the user's competence, intentions and needs (Ulich, 1985, 1987; Ackermann, 1987). Therefore we need to know (1) how users map the dialogue schema offered by the computer to an action sequence, and (2) how to capture the individual differences.

Computer science aspects

We need to (1) formulate requirements for the design of an adaptable interactive system and (2) investigate the consequences of different user interfaces (e.g. command structures) on the mental operational mapping process and hence on the outcome of the task-solving process.

We suggest an iterative approach to cognitive ergonomics, where the psychological investigation with the help of an adaptable system results in design requirements for user interfaces. These requirements in turn should serve as the input for further psychological testing.

2 EXPERIMENTAL STUDY 1: INDIVIDUAL OPTIMALITY OF DIALOGUE GRAMMARS

2.1 The Game PRIMP-1 ('Robi Otter')

To investigate mental aspects of task-solving processes, we have constructed the computer game PRIMP (*Programmable Robot for the Investigation of Mental Processes*). PRIMP allows the player to pilot a virtual robot in a maze displayed on the video screen (Figure 2). The robot's view is also

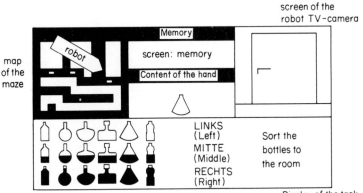

Basic commands

Go	The robot	moves one field forwards
Turn	The robot	turns 90 degrees to the left
Open	The robot	opens the door
Take	The robot	takes the bottle from the rack
Deposit	The robot to the empty rack	deposits the content of his hand
Remember	The bottle of the rack in sight will be represented in the memory on the screen	
Compare	The robot that represented in memory	compares the bottle in the rack with

The commands can be combined with an editor. There are the following possibilities:

until	door, wall, rack, free, empty, equal
if	door, wall, rack, free, empty, equal
not	

Example
Right = turn; turn; turn;
GT = Go until not free; Right; if door open;

Figure 2.   A section of the manual for PRIMP showing the basic commands and the screen showing the robot's view.

Command programs / Type of Command	AROBBY	B-BEFEHLE	D.BEF	ECMD2	GLOBI	HCOMMANDS
left		LEFT	L	L		
turn round		TURN ROUND	TURN ROUND			
right		RIGHT	R	R		
go ahead		GSTOP				
(Combinations)		RUN		GL/GR/VL/VR		
move left		WLEFT	RR (F)	XT		
recognizing Corners checking for Walls						
move right		WRIGHT				
recognizing Corners checking for Walls						
search empty Rack	DEPOSIT R		SLD +	PUT +	SKR R	PLACE L
search full Rack	SEARCH R			NEXT	SVR R	BOTTLE L
search Door	EXIT R		AUSGANG	TOR	ST R	EXIT L
bottle in Memory	GETBOTTLE R		GET	GET		
(+ with deposit/take) (L left/R right)						
from Room(x) to Room(y)		x.y (x,y E A,B,C)	Rxy (x,y E A,B,C)	xGOy	Rxy	
to Room(y)	RL/RM/RR					LEFT/RIGHT
from start to Room(y)	STARTRACK	START (nur R)	RSX	BEGIN (only R)	ROx	START (only R)
from Room(y) to Start		GOSTART	RXS	G8 (only R)		
Bottle in Memory autom.				GETL/GETM/GETR		BRING

Figure 3. Overview of the developed command programs (comparable commands are marked).

shown on the screen. The player's task is to sort bottles according to their contents. The robot can be directed with six basic commands which can be combined into iterations and selections. Moreover, commands can be embedded into macros which we call command programs. The names of the command programs can be chosen freely; they usually reflect the intentions and the goals derived by the player and indicate its scope of action.

In a first test series we could show that different cognitive styles—depending on abilities as on goal setting—lead to different 'command structures' and 'action sequences' (Ackermann, 1983, 1986a). In a second study six students developed command programs, tested the programs created by their colleagues, and reported their results (Ackermann, 1986b).

The command programs developed show differences in the use of the possible scope of action (Figure 3), thus reflecting individual strategies. Nevertheless, implementation of command programs comparable with respect to the offered objective scope of action can differ considerably (see command programs AROBBY and HCOMMANDS in Figure 3). The general objective in design of the command programs was to prevent own errors and errors of possible users. The other tendency was to find an optimal strategy. Figure 4 compares with the player's ranking when playing their own and the other command programs. In most of the cases, the efficiency is better when people are allowed to use their self-developed command programs.

| | Subject (Player) | | | | | |
	A	B	D	E	F	G
Parameter						
Time/Bottle	1	2	4	1	2	1
Commands/Bottle	1	6	3	1	2	1
Errors/Bottle	4	3	1	1	1	1

Rank orders of the individual developed commandsets in 6 trials (individual commandset and the best trial of each other)

Figure 4.   Players' ranking.

2.2   Analysis of the Mental–Operational Mapping Process with Prescribed Command Programs

In analysing the results reported we formulated the hypothesis that the developed command programs not only mirror the designer's action sequence but also represent which action sequences he wanted to propose to the user, thus defining the degrees of freedom in the user's scope of action. Furthermore, we assume that the mismatch between the designer's intentions and user's mental model is the main source of troubles in the human–computer interaction.

To test this hypothesis, we gave these command sets to other students without experience in creating command programs to see how prescribed command programs would be accepted and mentally represented and how this would influence their action sequence. The students were told to sketch their playing strategies on paper or to report them verbally. Figure 5 shows how differently the prescribed command programs were represented. While the drawing on the left is spatially oriented and based on the geography of the maze (in the sense of a cognitive map (Tolman, 1948)), the algorithm on the right shows mental representation of the command flow. Another interesting result is documented in Figure 6. The tendency to enlarge the scope of action prescribed by the command program according to individual needs (for example the need for control). The graph on the left is the designer's description of his command set and the drawing on the right is the user's interpretation of the offered scope of action. Curiously enough, this

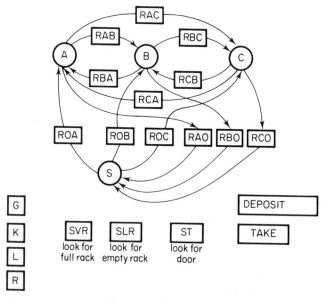

Figure 5. Different representations of the command program GLOBI (CAPITALS = defined Commands).

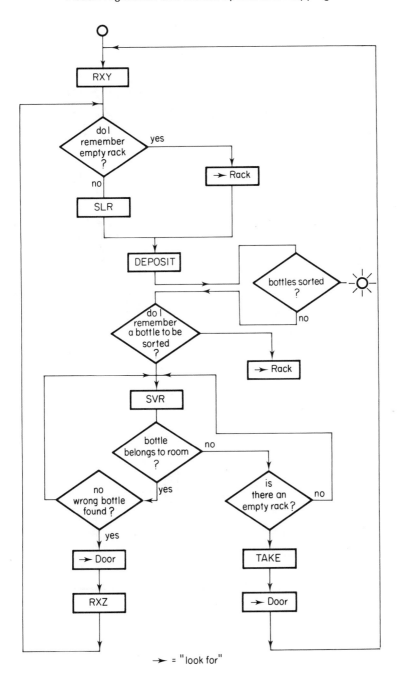

→ = "look for"

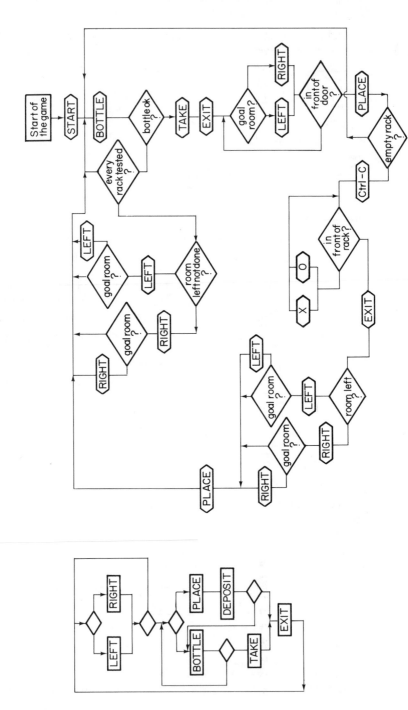

Figure 6. Intended (on the left) and subjective (on the right) scope of action of command program HCOMMANDS (CAPITALS = defined Commands).

tendency can be observed even when the given command program does not allow one to expand the scope of action or does not offer it explicitly. On the other hand, one player reduced the given command program which offers numerous possibilities to the commands he found necessary. Figure 7 shows its interpretation (Figure 3 includes the original command program, ECMD2).

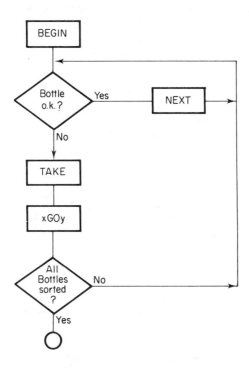

Figure 7.   Subjective scope of actions limited to the indispensable operations (CAPITALS = defined Commands).

Beside the discrepancies in the use of the possible scope of action, the verbal reports and sketched representations document individual differences in other aspects. The degree of abstraction, for instance, varies considerably. One player pointed out that the goal setting was his main interest and elaborated seven levels of goal settings reflecting the state of the game. Although one player employed all given command programs without noticeable difference in efficiency, several test persons refused to use some command programs (in particular, the command program AROBBY) or adapted or wrote their own versions that suited their intentions better.

2.3 Conclusions

The dialogue grammars offered by an interactive system are a subset of all possible operations and can often be combined to give more powerful commands, e.g. macros. The macros increase the level of abstraction and structure the dialogue grammar to suit the structure of the task better. It depends on the user, however, as to how he can make use of the offered dialogue grammar. The degrees of freedom of his scope of action are defined by the mental representation. The action sequences are the observable means that mirror the underlying mental representation.

The results presented show a wide range of different action sequences for equivalent command programs. We conclude that the task is represented in mental models which constitute different subjective scopes of action and vary individually in form and content. The analysis of the player's reports shows that the action sequences are influenced by cognitive styles, motivation, memory capacity and other individual skills.

3 EXPERIMENTAL STUDY 2: THE CORRESPONDENCE BETWEEN DIALOGUE AND ACTION SEQUENCES

In the experimental study 1, we showed that different cognitive strategies and command grammars can lead to improved efficiency in the task-solving process within the limited context of a game. Can these results be generalized as a guideline for a complex system with numerous different and more sophisticated interactive application programs? How does the opportunity to adapt the dialogue structure to the user's own preferences and abilities influence his interaction, efficiency and the quality of his work? Furthermore, can a designer offer such dialogue grammars as would induce a mental model that is more suitable for the given task, thus promoting the necessary learning effect? To investigate these questions we conducted a further series of experiments with the interactive system XS-2.

3.1 The XS-2 System

The Experimental System 2 (Stelovsky, 1985) defines a hierarchically ordered command language grammar based on regular expressions. This interactive system incorporates both the command language approach and menus (Stelovsky and Sugaya, 1985). The resulting command structures together with their syntax form command trees and are displayed as menus to the user (Figure 8). The user can conveniently adapt XS-2 different hardware settings or change the grammars that determine the structuring of his data (Stelovsky, 1984). Moreover, he can tailor the commands of an application according to his needs. A command can be renamed, the commands in a menu can be regrouped or subdivided into a menu hierarchy. Several alternative command structures for the same command can be introduced and tested within one application.

All these structural changes are done with one standard tool, which is also used to define the commands for a new application. The corresponding actions are simulated by an automatically generated program. This approach reduces the effort of application design: the commands for a new application can easily be developed and the existing commands can be reorganized without changing the application's implementation.

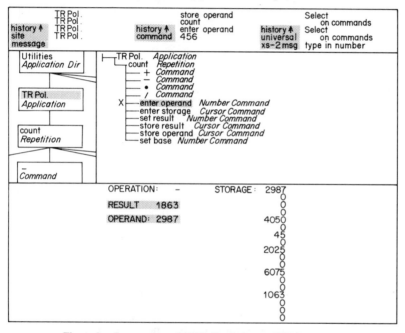

Figure 8.   Screen copy of 'TRPol', windows of XS-2 system.

3.2   A Pilot Study: Classification of Influence of Dialogue Grammars

3.2.1   Experimental design

To analyse and classify the influence of dialogue grammars on the action sequences, we made a pilot study with the XS-2 system. We selected three different tasks for a group of six students willing to participate in our study. The first task was to do some computations with a pocket calculator 'TRPol' (Figure 8). In the second task, they had to use the 'graphic editor' to draw a treble clef with the necessary lines. To design their own initials with the 'font editor' was the third task. They received one hour of introduction and a manual with a tutorial before they started work. Furthermore, they could use paper and pencil to document their thoughts and problems. The XS-2 system was enhanced by a logfile facility that enabled us to record all their actions.

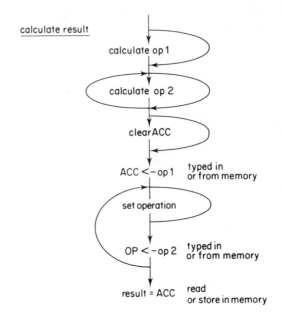

Figure 9.   Dialogue grammar of 'TRPol' according to the designer.

The calculator's dialogue grammar was aimed at simplifying continuous computations using the same operations (e.g. chain of additions). Figure 9 shows the designer's intention: the emphasized path shows that such chain calculations require only typing in numbers. Although the calculator could be used in the normal way, i.e. with the infix notation, the command structure did not explicitly offer this working style and rather suggested the use of the 'prefix polish notation'. This suggested dialogue schema was unfamiliar to our students who are used to commercial pocket calculators with either algebraic or the postfix polish notation.

The original command structure of the 'graphic editor' is presented in

Figure 10. The functionality of the editor had a limitation: once a sequence of lines and arcs was finished, it could not be modified. The drawing of such a sequence, however, could be aborted or the entire picture could be deleted. The command structure was intended to simplify a rather complex task: the drawing of a continuous chain of lines and arcs. The designer's intention led to an unfamiliar concept that cannot be found in a common graphic editor. Nevertheless, a single line, arc or circle could be drawn using the same command structure.

```
+ – – –  Draw Pictures                    "Application"
  + – – –  draw                           "Repetition"
    + – – – modifications                 "Repetition"
   I+ – – – insert graphics               "Command"
   I I+ – – – chain                       "Sequence"
   I I   + – – – starting point           "Cursor command"
   I I   + – – – continue with            "Repetition"
   I I      + – – – line ending at        "Cursor command"
   I I      + – – – arc                   "Sequence"
   I I         + – – – with centre at     "Cursor command"
   I I         + – – – ending at          "Cursor command"
   I I         + – – – orientation        "Selection"
   I I            + – – – as visible      "Command"
   I I            + – – – otherwise       "Command"
   I+ – – – delete all                    "Command"
  + – – – … other features …
```

Figure 10.   Command structure of the 'graphic editor'.

These dialogue grammars of 'TRPol' and 'graphic editor' are similar in their design philosophy: both command structures were intended to simplify the input of complex data objects (sequences of additions or lines). In contrast, it was not obvious how these concepts should be applied to define primitive objects (such as an operation on two numbers or drawing a circle). Indeed, both command structures were developed by the same person.

3.2.2 Results

In the calculator tool, the students tried to manage the offered dialogue schema according to their previous experience and its mental representation. Restructuring the given task to the unfamiliar suggested that prefix polish notation made more difficulties than the command activation itself. This led to a predominantly trial-and-error behaviour until the restrictions in the dialogue were learnt. Sometimes the calculator was operated using the not-obvious infix notation as a normal pocket calculator.

In the graphic editor, the first step—to open the appropriate data field—

was the most difficult aspect of the task. The correspondence of data to applicable commands caused some errors because the students tried to apply experiences or heuristics acquired with other software concepts. After they overcame this difficulty, no more dialogue-specific problems were observed. Few students had hardware-specific and individual-specific problems with the mouse. The action style changed from trial-and-error to goal-oriented command sequences.

The pilot study revealed that the influence of a prescribed dialogue grammar can be classified into task-specific, dialogue-specific and individual-specific aspects. Thus, it is important that, during the design of the user interfaces, a dialogue grammar can be tailored to suit a particular task, corrected to diminish the difficulties with the dialogue itself, and last but not least adapted with respect to the personal preferences.

3.3 Extended Study: Consequences of Different Command Grammars

In our further investigations with XS-2, we decided to focus on the task-specific, dialogue-specific and individual-specific aspects of the human–computer interaction and concentrate on how different dialogue grammars influence the user's actions. Therefore, a second pocket calculator 'TRNorm', which offers 'normal/algebraic' computations explicitly, was implemented. The tasks were redefined as well: instead of drawing clefs, the participants were asked to draw a car. Since everyone knows what a car looks like, our subjects should not have any task-specific problems. Another task was introduced: a simple table calculation with a newly implemented spreadsheet application.

3.3.1 Experimental design and research objectives

A total of 29 students participated in this series of experiments. They started with the conventional calculator 'TRNorm', changed to the drawing program and afterwards had to do a comparable calculation with the original calculator 'TRPol'. With the two calculators we offered different working environments with different command grammars to the user. They had to solve the same task—the sum of three easy computations—with both calculators. Our objective was to investigate how the action sequence varied depending on different working environments, and whether and to what degree people could adapt to prescribed dialogue grammars. The cars had first to be drawn with a single prescribed command structure.

Then, the participant's problems, expectations and needs were evaluated

TRNorm (left):

```
+ - - - TRNorm                        "Application"
+ - - - Compute                       "Repetition"
+ - - - set display                   "Selection"
I+ - - - new number                   "Number command"
I+ - - - from memory                  "Cursor command"
+ - - - +                             "Selection"
I+ - - - new number                   "Number command"
I+ - - - from memory                  "Cursor command"
+ - - - −                             "Selection"
I+ - - - new number                   "Number command"
I+ - - - from memory                  "Cursor command"
+ - - - *                             "Selection"
I+ - - - new number                   "Number command"
I+ - - - from memory                  "Cursor command"
+ - - - ⌐                             "Selection"
I+ - - - new number                   "Number command"
I+ - - - from memory                  "Cursor command"
+ - - - store display                 "Cursor command"
+ - - - set base                      "Number command"
```

TRPol (right):

```
+ - - - TRPol                         "Application"
+ - - - compute                       "Repetition"
+ - - - ±                             "Command"
+ - - - −                             "Command"
+ - - - *                             "Command"
+ - - - ⌐                             "Command"
+ - - - enter operand                 "Number command"
+ - - - fetch operand                 "Cursor command"
+ - - - enter display                 "Number command"
+ - - - store display                 "Cursor command"
+ - - - store operand                 "Cursor command"
+ - - - set base                      "Command"
```

Figure 11. Command structures of 'TRNorm' (left) and 'TRPol' (right).

and the 'graphic editor' augmented by additional command structures according to their suggestions. In a follow-up experiment, four students were asked to draw two cars. One car should be drawn using a new, sophisticated command structure proposed by an experienced XS-2 user. In contrast, they were free to choose among any of the commands in the enlarged set to draw the other car. Beside the interdependence between dialogue and action sequences, we could test how the variety of dialogue grammars met the requirements of differential design of interactive systems.

[1]   Computation-1 $\rightarrow$ Mem-1;
        Computation-2 $\rightarrow$ Mem-2;
        Computation-3 + Mem-1 + Mem-2 $\rightarrow$ Result

[2]   Computation-1 $\rightarrow$ Mem-1;
        Computation-2 + Mem-1 $\rightarrow$ Mem-1;
        Computation-3 + Mem-1 $\rightarrow$ Result

[3]   Computation-1 $\rightarrow$ Mem-1;
        Computation-2 $\rightarrow$ Mem-2;
        Computation-3 $\rightarrow$ Mem-3;
        [old] + Mem-1 + Mem-2 $\rightarrow$ Result

[4]   Computation-1 $\rightarrow$ Mem-1;
        Computation-2 $\rightarrow$ Mem-2;
        Computation-3 $\rightarrow$ Mem-3;
        Mem-1 + Mem-2 + Mem-3 $\rightarrow$ Result

[5]   other task structures

Figure 12.   Five basic strategies to solve the calculator task (the sum of three easy computations).

3.3.2   Results of the calculator study

As the entire sequence of user's actions was recorded on a logfile, we were able to reconstruct the motion on the command trees and derivate the underlying action sequences in the form of a set of rules. Figure 11 contrasts the command structure of 'TRNorm' on the left with that of 'TRPol' on the right. In Figure 12, the variety of different action sequences is generalized to five classes of working strategies. The classes are ordered according to their optimality with respect to the number of rules to be applied. For instance, applying strategy (1) needs less steps than (3) and (4). The results of 'TRNorm' and 'TRPol' are compared in Figure 12: 15 persons adhered to their computational strategy but 12 persons learned that they could improve their efficiency when they assimilated another strategy. To give an example, some participants took advantage of the fact that they can reuse the previous result in'TRPol'. In contrast to the pilot study, where numerous participants had dialogue-specific problems with the unfamiliar command structure of 'TRPol', only participant (V) had such problems. On the other hand, participant (L) restructured the task into chains of computations to make optimal use of the stored operation facility in 'TRPol' in the sense of the tool's designer.

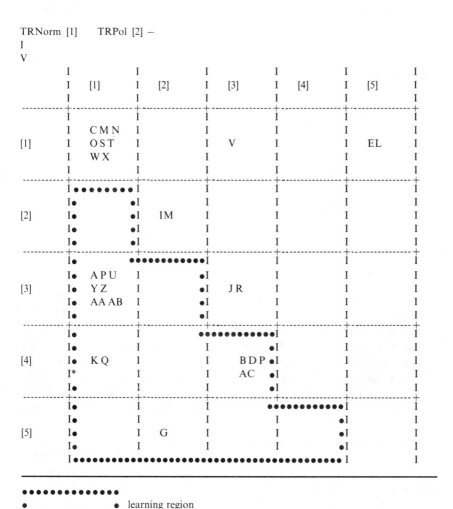

Figure 13.   Comparison of the basic strategies in 'TRNorm' and 'TRPol'.

3.3.3 Results of car drawing

The resulting cars were evaluated by a jury and the three with the best and the three with the worst rating are depicted in Figure 14 to demonstrate their aesthetic qualities. Certain cars look like professional sketches (AC), others rather resemble cans. Several participants tried to make the best out of lines and arcs that did not fit their intentions but some failed. The common explanation was a misleading expectation about what the commands can do. Analysing the process of drawing car (F), we can see how the participant managed to make the best of unintentionally drawn lines: 'dialogue-driven' is an apt word to characterize this drawing. The participant (F) erased the whole picture after the first steps and tried to redraw the car. But the dialogue-specific problems appeared again.

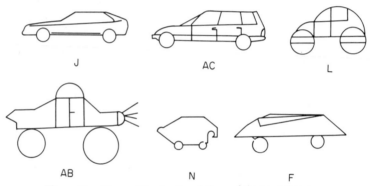

Figure 14.   Three of the 'best' and three of the 'worst' cars.

The comparison of the participants (F) and (AC) in the calculator and car drawing studies brings an interesting result. They used the same strategies in both calculators requiring unnecessary operations without any progress in learning (Figure 13). But in between the two computations, they had to draw the car, a task that was according to the jury's ranking solved best by (AC) and very inadequately by (F). To interpret this result, we interviewed (AC) and found out that he was very fond of drawing cars in childhood. Thus, his task-specific skills permitted him to assimilate easily an unfamiliar dialogue grammar.

In the follow-up study, the graphic editor was enlarged according to the users' suggestions by the command structures shown in Figure 15. On the left we see the main menu offering a variety of commands tailored for different subgoals (e.g. a single line or circle) as well as different strategies to define them. The differences in these strategies are mirrored in the parameter sequences (i.e. circle defined by centre first versus circle defined by the peripheral points first).

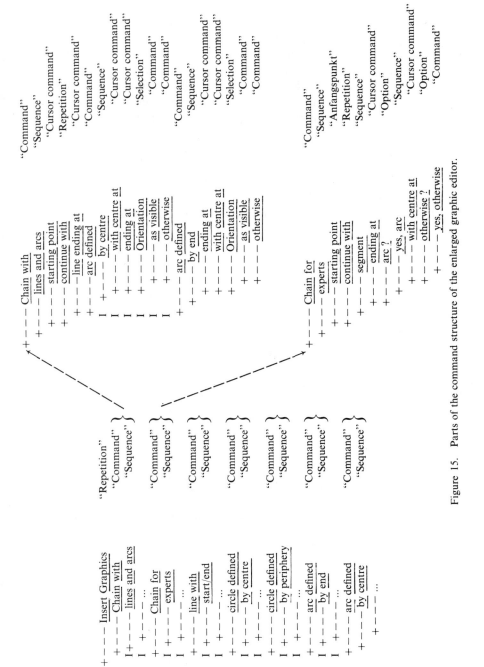

Figure 15. Parts of the command structure of the enlarged graphic editor.

Dialogue-specific problems occurred when participants had to use the prescribed command structure (e.g. person E) in Figure 16. When they were free to choose among commands, all of the participants were using varying command structures in correspondence with the subgoal and individual needs and skills. To give an example, wheels were drawn using one of the commands for circles. Interestingly enough, all participants were using one of the commands for drawing chains of lines and arcs when they found it most appropriate to accomplish a subgoal, e.g. when drawing the outline of the car's body. On the other hand, both such commands were employed according to individual preferences. The participant (AE), for instance, assimilated the sophisticated command structure and preferred it when drawing the second car—except for the wheels. Hence, we conclude that it is desirable to provide a larger set of alternative commands tailored to the task-specific requirements. Furthermore, an enlarged command set accommodates individual working style better and reduces dialogue-specific problems.

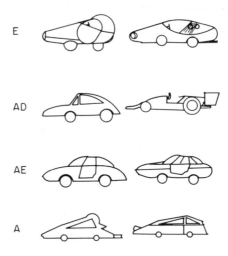

Figure 16.   Cars drawn with the enlarged graphic editor. Left: prescribed command set; right: enlarged command set with free choice.

4   CONCLUSIONS

In our experiments, we offered different dialogue grammars which structure the scope of action into different degrees of freedom. According to our results, it is impossible to induce only one action sequence even when the command grammar is very limited. The intended possibilities are redefined by the user. As the experiments with PRIMP-1 show, participants construct their own scope of action very individually to achieve the personally optimal action sequence, thus improving their efficiency. The resulting mental models of the task and the interaction are individually different in style, degree of abstraction and kind of representation.

These results are further confirmed by the calculator and the car drawing experiments. They demonstrate that the same task-solving process can be applied although the dialogue grammars differ. Conversely, one dialogue grammar will induce various action sequences, except for the trivial case where the dialogue grammar does not offer any degrees of freedom. The resulting action sequence depends on the individual user's mental processes. We try to outline the relations in Figure 17.

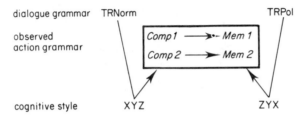

Figure 17.   The relations between dialogue grammar, action grammar and cognitive style.

In the car drawing experiment, it is interesting to see how certain participants manage to make the best out of unintentionally drawn lines. This shows the flexibility of the ongoing mental operational mapping process and adaptability to the restrictions imposed by the dialogue grammar. But the results depend on individual skills. Despite their efforts certain participants failed to reach their goal. Thus, the fact that a user's cognitive flexibility allows him to assimilate the prescribed dialogue is no proof of good dialogue design.

The classification and distinction of task-specific, dialogue-specific and individual-specific aspects of dialogue design proved useful not only in the analysis of the experimental results but also as a general tool for dialogue design. We can confirm Young's results (Young, 1981): different dialogue grammars are better suited for task-specific requirements. On the other hand, the comparison between the participants (F) and (AC) in the calculator and car drawing studies indicates that task-specific skills can also compensate for the dialogue-specific problems.

We can confirm our hypothesis, that problems in human–computer interaction are due to incompatibilities between mental operational mapping process and the offered dialogue grammar. On the other hand, different dialogue grammars can induce better strategies if they are more cognitively suitable with respect to the users individual mental-operational mapping process, thus confirming the principles of differential and dynamic design of production systems sensu (Ulich, 1978, 1987). Individual dialogue schemas adapted to the user's needs improve efficiency and prevent 'dialogue-driven' results. In Figure 18, we depict the interdependencies among the three aspects of dialogue design.

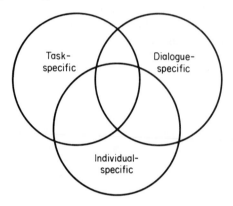

Figure 18.   The three aspects of dialogue design.

One of the most important results of our experiments is the influence of dialogue grammars on the learning process. As we could demonstrate in our calculator study and later confirm in the car drawing study, offering first familiar and simple dialogue grammars and then a large set of dialogue grammars with different degrees of abstraction and adherence to the task enlarges the subjective scope of action. The participants are motivated to explore the choices offered, learn the strategies suggested and then apply those that best fit their intentions. Such a stepwise learning process can even overcome dialogue-specific problems with very specialized dialogue grammars.

Hence, we conclude that an interactive tool should offer a variety of dialogue grammars to optimize the efficiency of a variety of users, tasks and tools. Therefore, we stress the importance of the adaptability in interactive systems. The example of XS-2 demonstrates the feasibility to construct such systems. Dialogue design is an iterative process, it should be dynamic with respect to the individual user and to his skills, preferences and needs. Such a design will support his learning thus improving his qualifications.

REFERENCES

Ackermann, D. (1983). Robi Otter oder die Suche nach dem operativen Abbildsystem. Interner Bericht der Studienarbeiten im SS. *Internal Report*, LAO/ETH Zürich.

Ackermann, D. (1984). Untersuchungen um Proyess der Handlungsregulation am Beispiel der Mensch–Computer-Interaktion. Interner Bericht der Studienarbeiten im SS. *Internal Report*, LAO/ETH Zürich.

Ackermann, D. (1986a). Untersuchungen zum individualisierten Computerdialogue: Einfluss des Operativen Abbildsystems auf Handlungs- und Gestaltungsspielraum und die Arbeitseffizienz. In: *Kognitive Aspekte der Mensch–Computer-Interaktion*. G. Dirlich, C. Feksa, U. Schwatlo and K. Wimmer (eds). Ergebnisse eines Workshops, 12/13 April 1984, Munich. Berlin: Springer.

Ackermann, D. (1986b) A pilot study on the effects of individualization in man–computer-interaction. *2nd IFAC/IFIP/IFORS/IEA Conference on Analysis, Design and Evaluation of man–machine-studies*. September 1985 (Varese). London: Pergamon.

Ackermann, D. 1987. Handlungsspielraum, Mentale Repräsentation und Handlungsregulation am Beispiel der Mensch-Comüuter-Interaction. Untersuchungen zum Prinzip der differentiellen und dynamischen Arbeitsgestaltung (Sensu Ulich). Doctoral dissertation, University of Bern, Switzerland.

Ackermann, D. and Stelovsky, J. (1987). The role of mental models in programming: From experiments to requirements for an interactive system. In: *Visual Aids in Programming* (Proceedings of an international workshop, May 1986, Schärding, Austria). M. Tauber and P. Gorny (eds). Berlin: Springer: Lecture Notes in Computer Science, 252, 53–69.

Cube, v. F. (1968). *Kybernetische Grundlagen des Lernens und des Lehrens*. Stuttgart: Klett.

Hacker, W. (1978). Allgemeine Arbeits- und ingenieurpsychologie. Bern: Huber.

Johnson-Laird, P. N. (1983). *Mental Models*. Cambridge, UK: Cambridge University Press.

Krause, B. (1982). Semantic information processing in cognitive processes. *Zeitschrift für Psychologie*, 190, 37–45.

Krause, W. (1982). Problemlösen–Stand und Perspektiven. *Zeitschrift für Psychologie*, 190, 18–36, 141–69.

Miller, G. A., Galanter, E. and Pribram, K. H. (1973). *Strategien des Handelns*. Stuttgart: Klett.

Moran, T. P. (1981) The command language grammar: A representation for the user interface of interactive computer systems. *International Journal of Man–Machine Studies*, 15, 3–50.

Oschanin, D. A. (1976). Dynamisches operatives Abbildsystem und konzeptionelles Modell. *Probleme und Ergebnisse der Psychologie*, 59, 37–48.

Reisner, P. (1981). Formal grammar and human factors design of an interactive graphics system. *IEEE Transactions on Software Engineering*. SE-7.

Schank, R. and Abelson, R. (1977). *Scripts, Plans, Goals and Understanding*. Hillsdale, NJ, USA: Erlbaum.

Spinas, Ph. (1986). VDU-work and user-friendly man–computer-interaction: Analysis of dialogue structures. In: *Psychological Aspects of the Technological and Organizational Change in Work*. L. Norros and M. Vartiainen (eds). Helsinki: Yliopistopaino.

Spinas, P. (1987). Arbeitspsychologische Aspekte der Benutzerfreundlichkeit von Bildschirmsystemen. Doctoral dissertation, University of Bern, Switzerland.

Stelovsky, J. (1984). XS-2: The User Interface of an Interactive System. Dissertation No. 7425 ETH, Zurich.

Stelovsky, J. (1985). User-tailored dialogue—Just a slogan? In: *Proceedings 8th International Computing Symposium*, March 1985, Florence. Amsterdam: North-Holland, pp. 345-52.

Stelovsky, J. and Sugaya, H. (1985). Command languages vs menus or both? *Software Ergonomie 85*. Stuttgart: Teubner, pp. 129–41.

Stelovsky, J., Nievergeit, J., Sugaya, H. and Biagione, B. (1985). Can operating system support consistent user dialogues? Experience with the prototype XS-2. In: *Proceedings ACM Annual Conference*. October 1985, Denver, pp. 476–83.

Tolman, E. C. (1948). Cognitive maps in rat and man. *Psychological Review*, 55, 189–208.

Triebe, J. K. (1980). Aspekte beruflichen Handelns und Lernens. Eine Feld- und Längsschnittuntersuchung zu ausgewählten Merkmalen der Struktur und Genese von Handlungsstrategien bei einer Montagetätigkeit. Unpublished Dissertation, University of Bern.

Ulich, E. (1978). Ueber das Prinzip der differentiellen Arbeitsgestaltung. *Industrielle Organisation*, 47, 281–6.

Ulich, E. (1985). Arbeitspsychologische Konzepte für Computerunterstützte Büroarbeit. *Spektrum* 14.

Ulich, E. (1987). Some aspects of user-oriented dialogue design. In: *System Design for Human Development and Productivity: Participation and Beyond*. P. Docherty, K. Fuchs-Kittowski, P. Kolm and L. Mathiassen (eds). Amsterdam: North-Holland, pp. 33–47.

Young, R. M. (1981). The machine inside the machine: user's models of pocket calculators. *International Journal of Man–Machine Studies*, 15, 51–85.

Zülch, G. and Starringer, M. (1984). Differentielle Arbeitsgestaltung in der Fertigung für elektronische Flachbaugruppen. *Zeitschrift für Arbeitswissenschaft*, 38, 211–16.

METAPHORS AND METACOMMUNICATION IN THE DEVELOPMENT OF MENTAL MODELS

Gerrit C. VAN DER VEER, Robert WIJK and Michael A. M. FELT

Department of Psychology, Free University, Amsterdam, Netherlands

In this chapter the use of computer systems or computer applications will be analysed from the viewpoint of human learning. This analysis is related to research being performed by an international working group supported by a grant from the Commission of the European Communities, action COST-11-ter. Previous analyses of human–computer communication and individual differences have been described by Van Muylwijk et al. (1984) and Van der Veer et al. (1985).

Any use of a computer presupposes a model of the system inside the mind of the user. The user may only plan his actions towards the system from this internal model, and any outcome of the interaction will be interpreted in relation to the model. A perceived deviation of the system's behaviour from the expectations will lead to an adaptation of the mental model. In order to start with a feasible mental model, adequate analogies and metaphors will have to be provided.

Apart from the effects of the communication between user and system concerning the delegation of the tasks the system is applied for, the system may communicate to the user about the interaction, in the sense of feedback, error messages and warnings. The user too may communicate, apart from task-related messages, about the interaction, about the communication languages and about problems in understanding the system's behaviour. These forms of communication, called metacommunication, will help the user to update his internal model of the system. A computer is very well equipped for metacommunication, and thus is well fit for learning and development of mental models, in parallel to performing the tasks the user delegates to it. The user interface will be described as the location of communication and metacommunication, providing a framework to separate the functions of the application system and the human–computer interaction component. Both the development of metaphors for the initial introduction of a system, and the application of metacommunication will be illustrated with examples.

COGNITIVE ERGONOMICS:
UNDERSTANDING, LEARNING AND DESIGNING
HUMAN–COMPUTER INTERACTION

1 THE ROLE OF MENTAL MODELS IN HUMAN–COMPUTER INTERACTION

When a human user has to interact with a computer system, he needs to develop a mental model of the system (Waern, 1987). This model is the source of the expectation the user has about the effects of his actions towards the system and the reaction of the system to his behaviour (Norman, 1983, 1986).

— It will guide his planning of the interaction.
— It will help the interpretation of the system's reactions.

An adequate mental model will be consistent with the user interface, or, to be more precise, with the conceptual model of the user interface as it is applied by the designer. The learning process that leads to a mental model, evidently has to start somewhere. Most of the time the system to be learned will be used for a kind of application, of which the user already has some prior knowledge, even if that may be of a non-computerized variant of the task domain. Building a new mental model for the computerized situation will be based on available knowledge of situations and systems. New models are built analogous to existing models. For the naive computer user the available 'old' models related to the user interface may involve:

1. *Human-to-human communication situations.* These processes may be used as analogies to either situations in which the system is 'only' an intermediary between different humans in communication on long distance or with time lags. Electronic mail systems and electronic conferencing systems are examples of such systems. Communication between humans may also be an analogy to verbal communication between a user and a system.
2. *Traditional work situations.* Many situations of humans performing a task in the traditional, non-computerized case, incorporate the use of tools, like paper and pencil, desktop calculators, filing cabinets and typewriters. The performance of tasks with the help of these tools will be a feasible source of analogies to understand the delegation of tasks to computer systems, e.g. in the case of spreadsheet systems, database management, or text processing.

The learning process may be facilitated by providing adequate metaphors. The non-computerized variants of work situations are 'natural' analogies for these processes and can be used as helpful metaphors. Human-to-human communications can, however, at the present state of the art rather serve as a misleading metaphor for human–computer communication.

When discussing the compatibility between user's and system's models it is useful to distinguish between different levels. Following Moran (1981), we have suggested that both the user's and the system's model can be described in terms of the following levels:

1. *Task level.* This level corresponds to the tasks which can be performed in the system. (A 'task' is intuitively defined as something which a user wants to achieve, and which is complex enough to require more than one simple action, cf. Card, Moran and Newell, 1983.)
2. *Semantic level.* The conceptual operations and conceptual objects as defined for the task by the system.
3. *Syntax level.* The commands available for accomplishing the operations defined at the semantic level and the wording of the feedback.
4. *Keystroke level.* The handling of the physical devices which are required to issue the commands, and the physical implementation of the feedback, as far as it is relevant to the human–computer interaction.

For the choice and construction of an adequate metaphor, both the level of the description of human–computer communication it represents and the mode of representation are important. Pictorial schemas and graphs are useful to represent structures and semantic relations, animations may illustrate processes, verbal descriptions are helpful to explain task analysis and task delegation. This choice of representation mode cannot be made without considering the levels of communication we derive from Moran's analysis. For task level and semantic level, metaphors may refer to known situations, systems and structures, or may be constructed by combining known concepts and schemes. For the syntax level, one will often need to construct a clear description that represents sequences, conditions, parameters. It may be better not to call these constructs a metaphor, but just a 'description' in the form of a BNF syntax, state transition diagrams, a set of production rules, or natural language. We will illustrate this with some representations that have proved useful in field research on introducing novices to computer applications in the domain of office systems.

Task level

For the introduction of an integrated office system (consisting of tools for database, spreadsheet, graphics, text processing, communication with data outside the system, and administration) verbal metaphors were developed (Van der Veer et al., 1987a). For two different kinds of users we constructed variant stories. For a group consisting of division managers, whose task was to consider the introduction of the integrated system into their division for utilization by their subordinates in order to receive reports as final results of the work of their group, the story was about the chief of an office who delegated the work to be done to a draughtsman, a calculator, a record-

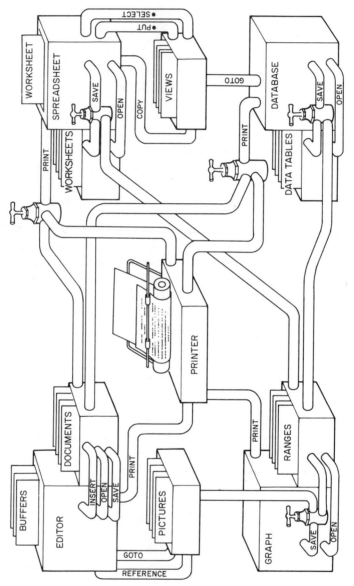

Figure 1. Visual–spatial metaphor of an integrated office system.

office, communication services, and a typing pool among other instances. For a group of employees who had as their main responsibility the administration of financial data structures, the story concerned the manipulation of a complex relational database, comprising tables about shop data, travelling salesmen records, merchandise statistics, and order forms.

Semantic level

For the structure of the semantics of the integrated system we just mentioned (the tools, objects and actions and their structural relations) a visual–spatial metaphor was developed as illustrated in Figure 1. The tools are represented by boxes, the objects by documents and the relations by tubes and taps, and the actions are denoted by their keywords, indicating the information flow produced by the command.

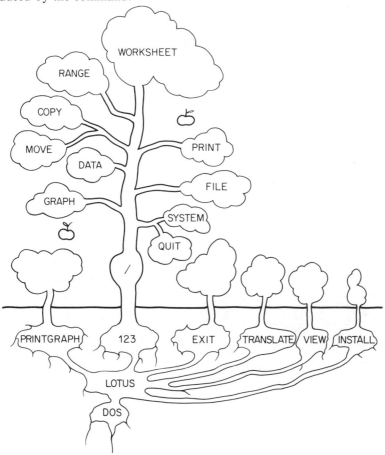

Figure 2.   Visual–spatial metaphor of a spreadsheet system.

The overall structure of the semantics of a spreadsheet system was illustrated with the help of another visual metaphor (Figure 2). The main tools in the system are represented by trees with a common root structure. The different facilities within the spreadsheet proper (the largest tree) are represented by main branches, grouped together on the basis of their semantic relationship. Information transfer between the facilities and between a facility of the spreadsheet and another tool are denoted by falling apples, indicating the direction of the information flow. The actions are denoted with their keywords, indicating the path in the tree structure they will activate.

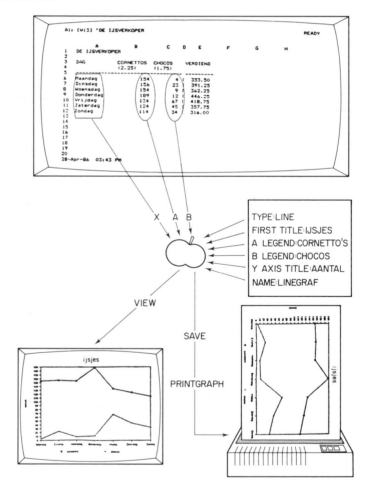

Figure 3.   Metaphor of graphic facilities.

Figure 3 shows the metaphors for the same system, illustrating some detailed semantic features of the graphic facilities. The attributes that are requested in the action of defining the graph are written alongside the icon we choose to represent the actual information, i.e. the apple. The possibilities to either ask for a preview at the screen, or for the storage and subsequent output on the printer, are represented with images indicating the actual orientation of the result.

The semantics of the command pair SAVE and RETRIEVE ask for the notion of the storage of an unknown number of different information packages, that may be retrieved in a random access mode. The typical fruit case illustrated in Figure 4 represents this type of storage.

FILE

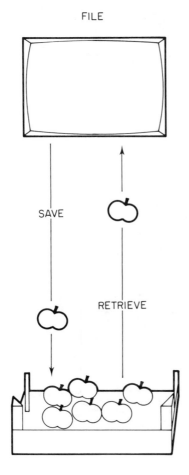

Figure 4.   Metaphor of random access storage.

Special attention was given to the semantics of the command PRINT. Apart from the 'natural' interpretation, illustrated with a denotation of the attributes of the information package written at the side of the icon for the information, another application of this command is in fact a special kind of storage action, on which occasion again the printing attributes are requested from the system. This connotated information is thereupon saved for future occasions, at which the Dutch proverb APPELTJE VOOR DE DORST ('an apple for a thirsty moment') hints, which, alas, has to be translated with the proverb 'a nest egg' (Figure 5).

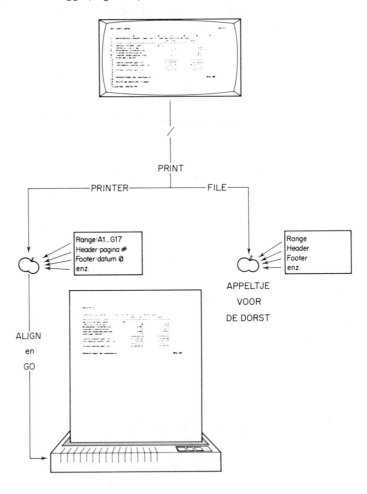

Figure 5. Semantics of command PRINT.

For the process that results from the execution of a macro in the spread-sheet system an animation was designed, from which Figure 6 gives a single view. In the actual animation, to be displayed on the PC, the caterpillar climbs the tree, and crosses the different elements of the macro (representing the series of command codes on the first line of the macro), until it reaches the end-of-command sign ('tilde'). The animal then drops from the end of the branch on to the worksheet, eats the content of a cell of the worksheet (as the last command of the series invokes the ERASE action), goes down one cell (illustrating the semantic meaning of the second line of the macro), and subsequently repeats the macro, since the last line in this example indicates the call of the same macro.

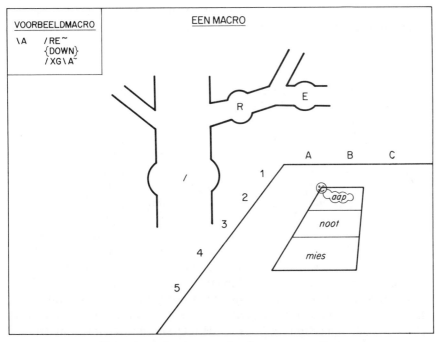

Figure 6.   Animation metaphor for macro execution.

Syntax level

Representation at this level often is closely related with the actual appear-ance and physical characteristics of the interface, e.g. the layout of the screen. For systems with alternative ways of interaction this may lead to contradictory descriptions. In the spreadsheet system mentioned before, the beginner user will be happy to let his interaction be guided by a menu, displayed on top of the screen. The line immediately below the line of menu options most of the time shows the submenu that will be offered if the option is chosen that is momentary blinking (an action of attention of the menu cursor). For novices this leads very often 'naturally' to a syntactic model of a top-down tree of menu choices. After considerable training (e.g. 25 hours) most users turn to the possibility of manual input of the first characters of these choices. This results in the behaviour of defining a command string consisting of sequence-dependent single characters, keyed in as if typing from left to right. This is (at least sometimes) accompanied with an internal model of the syntax consisting of the commands (or their first character) structured in a horizontal tree, top-left. Offering novices one of these representations on paper will either support the original 'natural' mental representation or the one that may be found after some experience with the system. In any case, a correct representation of the command structure seems (in a pilot study) to be related to spatial ability, a result that is in agreement with some results Rothkopf (1986) reports.

2. COMPUTER SYSTEMS AS ADAPTIVE ENVIRONMENTS: METACOMMUNICATION

Computer systems are a kind of environment optimally fit for stimulating the learning process. Interaction about the task delegation by the user to the system may be accompanied by communication about this first type of interaction. This second type of interaction or communication we call metacommunication. The effect of metacommunication will be the adjustment of the mental model the user applies for planning and interpretation. Metacommunication may be incorporated in the user interface, both in explicit form (help facilities, error messages, coaching), and in implicit form (command names, icons and screen layout may suggest the meaning of elements and structure of the human–computer communication). This second type of metacommunication seems not always to be recognized by designers of user interfaces. In section 2.2 some examples will be presented. The metacommunication facilities in the interface may adapt to individual differences between users, and to individual needs for the improvement of the mental model. Some users tend to have quite extreme needs for validation, feedback, or online information, either at the start of the learning process, or for prolonged periods during practice and gaining expertise. Some will request coaching, others might ask for examples of correct application. Computer systems are well suited to this extra task, having large resources for responsiveness (real-time processing, multimode communication, large storage capacities).

2.1 Explicit Metacommunication

Explicit metacommunication is designed especially to serve the user. The main purpose of information exchange between the system and the user is the clarification of the task-related interaction, the provision of extra information about the way of delegation of tasks, and the evaluation of effects of user behaviour and system reactions. It may be user-controlled, as in situations where the user asks for help, examples, or special information. Alternatively the system may start the metacommunication, either on the basis of any kind or model of the user, incorporated in the user interface, or at the occasion of an uninterpretable, ambiguous or erratic user message.

2.1.1 User-controlled metacommunication

This way of explicit metacommunication is initiated by the user. Apparently he experiences a momentary need for clarification, either because his mental model is perceived to be incorrect or incomplete, or he is uncertain about his skills or his interpretations of the situation. In existing systems the possibilities to acquire the kind of information needed are often limited. The user often lacks knowledge at one level of the interaction. For example, at the semantic level when he wants to accomplish a certain task, but he may not know (or does not remember) which objects, tools or actions are available for this purpose or which attributes apply. In that case he will be unable to ask for help indicating the relevant command name (as for example is necessary for the consultation of the UNIX online manual). A help system should allow him to search for solutions triggered by the name of the subtask. As the task description hardly ever leads to one single terminology, the possibility to choose from a menu will be the best opportunity.

In case the user knows enough about the semantics of the system to choose the relevant command, on the other hand, only syntactic help will have to be presented. The designer of the presentation mode will have to take into account that many users of today's application systems are not professional computer scientists. Expressions derived from operating systems' 'computerese' are not very self-explanatory. For some types of users (so called 'imagers') a graphical representation will be an improvement, as for others a well-chosen set of examples may be required.

Users show individual differences in their strategies of exploring a new system. Some only feel safe if they have browsed through a lot of information about the system, building their mental model during this process, to compare it with reality only afterwards. Others like to learn by doing, asking for help only at the occasion of an actual uncertainty in the application of their mental model. Adaptation to this variation in needs is possible if all kinds of help are available in all situations and modes.

A special kind of metacommunication that is available in some educational situations, results from the presence of an online coach system. Experiments in our laboratory (Beishuizen, 1986) have been conducted to investigate strategy differences between students searching a database (briefly to be described as serialistic vs holistic strategies). It was shown that coaching novices to be consistent with the system's model of the user's individual strategy leads to an improvement in performance (if this model of the user was based on prior behaviour of the user on related search tasks). On the other hand, the user should always be offered the possibility to ignore the guidance and stick to his own momentary strategy, and even to silence the coach. If this last possibility is not available, coaching is a case of system-controlled metacommunication, and the user is forced to build a mental model of the interface which fits the instructor. We do not advocate such an interface for adult novices.

2.1.2 System-controlled metacommunication

Metacommunication initiated by the system nearly always occurs if the system either cannot decide how to interpret the user's behaviour on a syntactic level, or in case the result of a system action (invoked by the user's behaviour) has produced disaster. Error messages should clearly indicate the syntactic incompleteness that is probably the reason in the former case, or point to semantic inconsistency that in most situations is the cause of errors of the second type. Moreover, in the last case the user should be offered the possibility to undo the effects of the accident.

System-controlled metacommunication may also occur as an indication of the completion of a command, possibly combined with feedback about the effect. This will give the user the possibility to check whether his purposes have indeed been accomplished, hopefully again with the opportunity to undo the act. Silent systems like the UNIX shell deprive users of this facility, which all too often results in the loss of data or work.

2.2 Implicit Metacommunication

Even when the designer of the human–computer interaction did not think of metacommunication as a goal, it will take place. Utterances of the system, and user actions with their side-effects, will lead to perceptible changes in the situation. These effects, if attended to by the user, may lead to validation of his mental model. Interpreted in the light of his existing mental model, the perception of system reactions and their side-effects will result in certain expectations of future system reaction, or in decomposition of tasks into subtasks and individual actions. Alternatively, if the perceptions are inconsistent with the actual mental model, this inconsistency motivates an adjustment of the mental model.

2.2.1 Verbal representation

Implicit metacommunication will be found in the naming of commands. If names are not chosen with the purpose of metacommunication, undesirable side-effects may result. For example, some unexpected terminology is to be found in the command language of a couple of British database systems that are explicitly designed for educational purposes (Freeman and Levett, 1985; GRASS, 1985). In one of these systems, the command to select fields (columns) in a data table for further processing is PRINT, while the command for printing the ultimate selection is OUTFILE__LST. In the other system, a semantic peculiarity is constructed, in the sense that for calculating the arithmetic mean of a set of numbers, one has first to choose a tool called GRAPHS, whereupon the option AVERAGE becomes available. Another example of a bad instance of implicit metacommunication is the option D in UNIX MAIL, which invokes the mnemonic 'delete' (which is, in the explicit metacommunication of the online manual, described in this way). An attentive user will hopefully discover that mail attributed D is not deleted, but only appointed 'TO__BE__DELETED'. The object remains available as long as the user stays within the system. Only when he exits the system, will a real deletion take place, and then only if some particular ways of exiting are used, intentionally or unintentionally. The facility is not quit, and in the case of other legal ways to exit the facility no deletion will take place.

2.2.2 Graphical representation

Recent developments in the use of graphics have already led to some improvements. The application of icons instead of character strings to indicate commands or objects increases the possibility to offer relevant implicit metacommunication. Icons may include symbolic representation of the combined semantic information about objects, location, action and/or direction. Figure 7 presents an example from the Heidelberg Icon Set (Chang et al., 1987), representing the commands *insert line*, *insert string*, *delete string*, *move string*, and *replace string*.

It includes the notion for action INSERT (the arrow), objects TO__BE__ INSERTED (the square on the right) and INSERTED__IN (the two squares on the left), and location AFTER__PLACE (the shaded part of the indicated object) whereas the usual verbal command is INSERT or even I. The choice from verbal menus instead of the application of commands often does not offer much more indication of semantics. However, an iconic interface (even combined with direct manipulation) is no guarantee either for correct implicit metacommunication. Tauber (personal communication) gives the example of the Macintosh clipboard, which is an inadequate metaphor because the content is not 'unclipped' when pasted in some place, only 'copied from'. This is inconsistent with the storage mechanism of the trash can, that indeed is empty after its content—always only the last deposit—has been re-used.

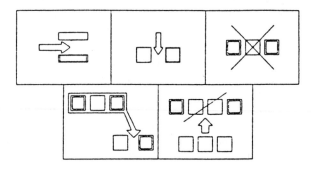

Figure 7.   Examples from the Heidelberg Icon Set.

The use of graphical programming tools like online construction of structured diagrams for control flow in programs offers another possibility to communicate the semantic meaning of the strucutral elements. Van de Veer et al. (1987b) showed that this tool enables novice programmers to learn, even if they lack the mathematical background that in other situations turned out to be a bottleneck in mastering a programming language. Graphic representation of control structures makes it easier to control the scope of iterations and their termination, and to preserve a mental model of the structure of algorithm under construction.

3 THE USER INTERFACE AS A LOCATION OF METACOMMUNICATION

As far as metacommunication facilities are incorporated in the system, they are located in the user interface. This may be considered as a layer between the application system and the user (Tauber, 1986). In designing the user interface, the knowledge of human learning will have to be applied to make adequate use of the possibilities of adaptation and flexibility that are available.

3.1 Individual Differences

The conceptual model of the user interface has to be developed with the explicit aim to be parallelled by mental models. Mental models of users are individually different representations of the system by different users, all of which will start by being a novice to this system.

Amount of experience
As we know that novices lack the availability of sufficient memory support for remembering the names and syntax of commands, they should be allowed to choose from (semantically well structured) menus or to work in a form-filling mode. After some hours of experience, they will need the possibility to turn to command mode. If novices are not sure about semantics, well-designed icons may be helpful.

Imagers/non-imagers
Studies on the difference between 'imagers' and 'non-imagers' (Van der Veer et al., 1987a) revealed that the former construct their mental model according to verbal or visual–spatial metaphors presented during tutorial training. The other type of user needs a lot of hands on experience with exemplary cases, which could be provided inside the user interface in a special 'example'-mode, that guards the novice from disasters during the first phase of development of his mental model.

Occasional users
Occasional users of a system may in certain instances reach a situation in which they perceive their mental model to be inadequate. In this situation they need the possibility to ask for information about the present mode of interaction, about previous situations and actions, and about possibilities to proceed. These kinds of information should always be immediately and overtly available, e.g. by pressing explicitly marked keys.

3.2 Attention

When the designer of the user interface intends to draw the attention of the user towards relevant information, he needs to make adequate use of knowledge of human attention span. Too many colours, too many windows or overlays on the 'desktop', or too many different character sizes or cursor types only distract the user. These features will actually hide the relevant information that has to be related to the (fragmentary) mental model of the user, especially as his model may lack some of the slots, for which the relevant information provides the keys.

3.3 Consistency

A last aspect of the relation between the conceptual model of the user interface and the use of metaphors is the internal consistency of the user interface. In some commercially available systems, the user interface shows certain characteristics which are really inconsistent. This prevents the user from constructing a comprehensive mental model. On this occasion we may choose to present a different 'Ideal' conceptual model to the user, with the help of an elegant metaphor. Upon explanation of the metaphor and its relation to the user interface, we may tell the novice that the user interface deviates from the conceptual model in certain circumscribed details. This helps the user to stick to a simple mental model, at the same time enabling him to cope with the exceptions.

REFERENCES

Beishuizen, J. J. (1986). Leren opzoeken van informatie (Learning to search for information). The Hague: Stichting voor Onderzoek van het Onderwijs.

Card, S. K., Moran, T. P. and Newell, A. (1983). *The Psychology of Human–Computer Interaction.* Hillsdale, NJ, USA: Erlbaum.

Chang, S. K., Tauber, M. J., Yu, B. and Yu, J. S. (1987). The silicon compiler—an icon-oriented system generator. In: *IEEE Workshop on Languages for Automation.* Washington, DC, USA: Computer Society Press of IEEE.

Freeman, D. and Levett, J. (1985). Quest in the learning environment. In: *Proceedings of the 4th World Conference on Computers in Education.* WCCE/85. Amsterdam: North Holland.

GRASS (1985). Graphics Searching and Sorting. An information handling package from Newman College. Birmingham: Newman College.

Moran, T. P. (1981). The command language grammar: a representation for the user interface of interactive computer systems. *International Journal of Man–Machine Studies,* 15, 3–50.

Van Muylwijk, B., Van der Veer, G. C. and Waern, Y. (1984). On the implications of user variability in open systems—An overview of the little we know and of the lot we have to find out. *Behaviour and Information Technology,* 2, 313–36.

Norman, D. A. (1983). Some observations on mental models. In: *Mental Models.* D. Gentner and A. L. Stevens (eds). Hillsdale, NJ, USA: Erlbaum.

Norman, D. A. (1986). Cognitive engineering. In: *User Centered Design.* D. A. Norman and S. W. Draper (eds). Hillsdale, NJ, USA: Erlbaum.

Rothkopf, E. Z. (1986). Machine adaptation to psychological differences among users in instructive information exchanges with computers. In: *Man–Computer Interaction Research —MACINTER I*. F. Klix and H. Wandke (eds). Amsterdam: North-Holland.

Tauber, M. J. (1986). An approach to metacommunication in human–computer interaction. In: *Man–Computer Interaction Research—MACINTER I*. F. Klix and H. Wandke (eds). Amsterdam: North-Holland.

Van der Veer, G. C., Tauber, M. J., Waern, Y. and Van Muylwijk, B. (1985). On the interaction between system and user characteristics. *Behaviour and Information Technology*, 4, 284–308.

Van der Veer, G. C., Felt, M. A. M., Van Muylwijk, B. and Van Biene, R. J. (1987a). Learning an office system—A field study on the development mental models. *Zeitschrift für Psychologie*, suppl. 9, 46–58.

Van der Veer, G. C., Van Beek, F. and Cruts, G. A. N. (1987b). Learning structured diagrams—effect of mathematical background, instruction, and problem semantics. In: *Visualization in Programming*. P. Gorny and M. J. Tauber (eds). Heidelberg: Springer.

Waern, Y. (1987). Mental models in learning computerized tasks. In: *Psychological Issues of Human Computer Interaction in the Work Place*. M. Frese, E. Ulich, and W. Dzida (eds). Amsterdam: Elsevier.

AN INVESTIGATION OF THE LEARNING OF A COMPUTER SYSTEM

Michael WILSON,* Philip BARNARD** and Allan MACLEAN†

Rutherford Appleton Laboratory, Chilton, Oxon, UK
MRC Applied Psychology Unit, Cambridge, UK
Rank Xerox Ltd, EuroPARC, Cambridge, UK

1 INTRODUCTION

It is widely acknowledged that the behaviour of users and their underlying cognition is complex. For various purposes, it may be necessary to provide a detailed characterization of user behaviour and cognition which is unlikely to be derivable from a single theory or assessment technique. Consequently, a number of specific assessment techniques and theoretical interpretations must be matched to specific purposes. To match these two, it is necessary to understand the properties of the possible theoretical interpretations and assessment techniques.

Assessment techniques can be used either for research purposes, to provide information about how users represent and process information, or to assess commercial products as part of the design process. In the design process, this assessment can be at an early stage, in order to provide ideas, or late, in order to locate user difficulties. If one wishes to obtain quick results with minimal analysis effort in the design process there must be a conscious trade-off between the effort to complete any test and the speed and amount of information provided. Therefore, it may not be feasible to apply the full range of detailed and laborious measures used for research purposes. Nevertheless, the interpretation of research tests and the relationships between them can provide criteria for selecting which tests may be appropriate, and their validity for the more practical problems of designing and assessing a system.

This chapter considers different assessment techniques in the context of a study of users learning a commercial integrated office system. The purpose of the study was to trace the development of different aspects of the

COGNITIVE ERGONOMICS:
UNDERSTANDING, LEARNING AND DESIGNING
HUMAN–COMPUTER INTERACTION

cognitive processes and representations which support user performance during the learning of a system. Since it is acknowledged that no single assessment technique could deliver this, several different techniques were used. To incorporate the results delivered by these different techniques, it is necessary to develop explicit models of what each test delivers, and the view of the user on to which such results will be mapped. An attempt will be made in this chapter to be explicit about both.

The initial view taken of user learning is characterizable by a general model of skill acquisition (after Fitts, 1964). This model divides learning into three phases. In the first, users acquire sufficient fragments of knowledge about a system to support the performance of some tasks. In the second phase, the knowledge recruited to perform tasks is 'compiled' into procedures. In the third phase, users draw on these compiled procedures and exhibit performance which can be considered 'expert like'. A further distinction exists in this model between the accessibility of the representations of knowledge in the first and third phases. It is assumed (after Anderson, 1983) that the knowledge in the first phase is accessible to the processes of verbalization, whereas the compiled procedures drawn on in the third phase would not be recruitable. When task performance is described by users who have reached the third phase of learning for those tasks, the explanation will not be based on the 'compiled' knowledge which actually supports performance, but on other knowledge that they hold about the system. This knowledge may be either consistent or inconsistent with the knowledge compiled into procedures, but may be drawn on during the performance of novel tasks which are not proceduralized.

The assessment techniques used in this study embody three basic strands: performance measures; measure which allow access to articulatable knowledge; and measures which implicitly require subjects to use knowledge about a system. There were overlaps and differences in the aspects of performance and cognition captured by the tests used in the study. This chapter will outline the techniques used, and illustrate both the similarities and distinctions in the results, in order to understand what aspects of cognition and behaviour each test is best at capturing. In particular, the view of learning which can be captured by these tests will be described.

2   DETAILS OF THE STUDY

In the study a sample of eight naive computer users learned how to use the VisiOn (trademark of VisiCorp) interactive office system. This system incorporates three principal environments for word processing, graph drawing and spreadsheet calculation. These environments have a common conceptual interface in which users invoke command operations by pointing with a mouse at command words in linear menus. Once a command is invoked users follow prompts for the required mouse and keyboard actions (see Lemmons, 1983, for further details of the VisiOn system).

The study consisted of eight sessions which were systematically structured to enable the training, exercising and testing of a broad sample of VisiOn functionality. The overall design of the study is presented schematically in Figure 1.

SESSION

1	TUTORIAL	VIDEO PRACT A1		
2	VIDEO PRACT A2	A SET		
3	VIDEO PRACT B1	A SET		PKE
4	VIDEO PRACT B2	A SET	B SET 1	
5	VIDEO PRACT B3	A SET	B SET 2	
6	A SET	B SET 3	C SET	PKE
7	RECALL, QUESTIONNAIRE, DATA ENTRY TEST			
8	DATA ENTRY TEST			

Key: VIDEO PRACT—watching a video of command sequences, imitating them, and then practising each command sequence. A1 and A2 for A set test. B1 for B set 1, etc. A, B and C SETS—three performance tests. PKE—Prompted Knowledge Elicitation test. RECALL— A command name sorting test.

Figure 1.   The structure of the testing sessions in the observational study on VisiOn.

In session one, users familarized themselves with the keyboard and mouse as well as the basic concepts associated with windows and menus by using the inbuilt tutorial within the VisiOn system. Users then began a systematic training programme. The training was by example and illustration. Users were shown a videotaped example of the method for each task, and asked to perform the task they had seen using the same materials as in the videotape during which time they could ask the experimenter about any details of which they were unclear. After a block of three tasks had been watched and copied, subjects practised each task three times on new materials. Materials were presented to subjects as manuscripts with editing changes marked on them. As the sessions progressed more and more functions were illustrated, copied, and practised.

There were four different performance measures (called the 'A', 'B' and 'C' set tests, and a data entry test). Although these included subjects verbalizing occasionally, there was one explicit test of verbalizable knowledge: the Prompted Knowledge Elicitation test (PKE). There were two tests that implicitly required the use of the subjects' knowledge of the system, a questionnaire study and a task which involved the sorting of command terms into the system hierarchy. Details of individual tests are given later.

The training was arranged with sessions 1 and 2 on consecutive days and the remainder at intervals of 3 to 7 days. The duration of each session was fixed at 2 hours since this was thought to be a reasonable sample period for the intended users of such systems.

3 PERFORMANCE TESTS

Each of the performance tests in this study was selected to answer specific questions about the process of learning. Firstly, how does performance improve with practice and is there any evidence of the proceduralization expected as a result of the general model of expertise? Secondly, do users incorporate new functions more easily as a result of the accrual of general system knowledge? Thirdly, how does this knowledge generalize to untaught functions? Fourthly, when users reach a state of expertise in a specific task, do they make judgements as to the most appropriate method to perform that task on the basis of the time to use the possible methods, or on the basis of other knowledge?

The general view of learning outlined above suggests that the information that supports behaviour passes from being declaratively represented to being procedurally represented. During this change in the representation of knowledge, there is an hypothesized parallel reduction in the time to perform the task and the number of errors during performance.

This change in performance time has been formalized in the 'power law of practice' which states that the time to perform a task decreases as a power-law function of the number of times the task has been performed. It has recently been argued that this law not only applies to the domain of motor skills where it was originally recognized (Snoddy, 1926), but to a full range of human tasks (Newell and Rosenbloom, 1981), including perceptual tasks such as target detection (Neisser et al., 1963) and purely cognitive tasks such as supplying justifications for geometric proofs (Neves and Anderson, 1981). Consequently, it is often assumed that as experience with a computer system increases, the general task performance time and error count will reduce.

The A set test was designed to assess the general reduction in performance time and errors for commands used frequently. A detailed report of this test is given elsewhere (Wilson et al., 1985a, 1985b), so only a brief summary will be presented here. In the A set test, subjects first imitated and then practised nine basic command sequences during the first two sessions, and then used these commands in later sessions. Three command sequences were selected from each of the three environments in the VisiOn system. The commands were: for word processing—delete a word, move a block of text, enter an item of text; for spreadsheet use—enter a number, enter a label, enter a subtraction or summation formula; for graph plotting—select data for plotting, select a style for plotting (a line graph), plot a graph. Subjects were tested on the performance of these nine commands on sessions 2 to 6 (but using different materials), by being presented with manuscript pages compatible with each of the three environments, marked with editing instructions which required the use of the A set command sequences. The materials were similar to those which the subjects had used to practise the commands. This procedure resulted in five progressive samples of performance on each of the nine frequently used commands.

Anecdotal evidence and personal experience suggest that a division is made by users within the command set of a system, between the subset of commands which are used regularly and the 'other' commands. Roberts and Moran (1983) have defined a set of core commands for text editing systems which encompass the functionality required to produce any real output. This set is comparable with the subset of system commands which most users regularly apply (e.g. 'delete text', 'enter text'). In contrast, there are many commands available on computer systems which even experienced users do not use, and of whose details they are often uncertain. These may not be learned until some advanced task must be performed on a system. For example: a user may not use the commands to create a contents page or an index in a text processing system until producing a book, although considerable experience of the system may have been gained writing letters and memoranda.

The commands acquired early in system use will be learned at a stage when there is little knowledge of the system which could either aid in that learning or interfere with it. The commands acquired later during system use will be learned when there is a larger body of knowledge held about the system. This knowledge should benefit the later learning of commands if they are consistent with it, but may inhibit learning if they are incompatible.

The development in the range of commands used has been actively embodied in the 'training wheels' approach to learning computer systems (Carroll and Carrithers, 1984) where users are initially exposed only to the core of commands required to fulfil their task. Users are exposed to other commands only after they have attained some proficiency with the core commands. This technique has been shown to lead subjects to perform basic tasks more quickly than subjects who use the whole system initially (Carroll and Carrithers, 1984; Catrambone and Carroll, 1987). This approach can be integrated with the initial view of the user outlined above through the 'chunking hypothesis' and three assumptions which support it (after Rosenbloom and Newell, 1987):

The chunking hypothesis. A human acquires and organizes knowledge of the environment by forming and storing expressions, called *chunks*, which are structured collections of the chunks existing at the time of learning.
Performance assumption. The performance program of any system is coded in terms of high-level chunks, with the time to process a chunk being less than the time to process its constituent chunks.
Learning assumption. Chunks are learned at a constant rate on average from the relevant patterns of stimuli and responses that occur in the specific environments experienced.
Task structure assumption. The probability of recurrence of an environmental pattern decreases as the pattern size increases.

This hypothesis suggests that the components of command sequences will be progressively clumped together into chunks until a whole command sequence is represented as a single chunk. The fewer chunks that are required to perform a command sequence then the less time the sequence will take to perform. Consequently, commands learned at an early stage will have no or few relevant constituent chunks present and will require the development of a procedure for the sequence from the minimal parts of the sequence. The representation of commands learned later may include chunks which were developed for commands already learned. However, they will still require the development of some structures which were not previously defined and the appropriate recruitment of those which are. This process may be easier if the new commands conform to a characterization of the commands already learned (Barnard et al., 1981). In contrast to chunks of performance sequences, such characterizations may take the form of high-level rules governing the system command structure and operation which users may have abstracted. The performance tests were designed to capture this hypothesized variation between commands learned early in system experience and frequently used, and those learned later and used more rarely.

Whereas the A set tested core commands of the class described by Roberts and Moran (1983), the B set performance test was designed to assess the acquisition of new functions as learning progressed and to assess whether users incorporate new functions more easily as a result of the accrual of general system knowledge. To do this, command sequences learned in one session were tested in a subsequent session. However, in each case the command sequences were tested only once. In tests from sessions 4 to 6, 18 different command sequences were tested. The command sequences required to perform the B set tasks incorporated sections of the command sequences required for the A set tasks. For example, the command sequence for 'move text' in the A set required the menu selection of 'move' and then a selection of the area of text to be moved and the target point; the B set command sequence for 'copy text' required the selection of the menu item 'copy' and then the same selections to define the body of text and target site. In this way, the B set command sequences incorporated chunks which had already been established for A set commands. If the chunking hypothesis and its associated assumptions hold true for the learning of computer command sequences, then the B set commands should be learned more easily than those for the A set since chunks intermediate between the single actions and the full command sequence should already exist.

To decide if the knowledge acquired generalizes to untaught functions, a third performance test (the C set) was administered to capture expert performance on commands unknown to the expert. This test was administered at the end of session 6, and involved users attempting six tasks which required the use of novel system functions. The command sequences for these functions were not explained to the subjects, although they required the selection of menu items which appeared on menus with which subjects were familiar from their general use. This test was intended to show the extent to which general system knowledge could be recruited to perform the high-level task of discovering the appropriate command sequence for novel functions. However, the test was only attempted by the three fastest subjects, who made only five errors between them. These are too little data on which to answer the question conclusively, although this was a reasonable indication of the generalization from knowledge of other parts of the system, for at least the fastest subjects.

The general model of learning described above does not characterize expert performance. Some psychological models of expertise (e.g. Card et al., 1983) have assumed that expert users make decisions as to the optimal method to perform, only on the basis of the time to complete a method. This expression may be too simple to explain performance and other factors may have to be taken into account. To assess if other factors are involved in the decision, the fourth performance test was designed. A data entry test was used to sample the choices made by experts between methods to achieve tasks. The data entry test was specifically focused on the trade-off between two methods of entering data into a spreadsheet (this study is reported in detail elsewhere: MacLean et al., 1985). This test took place in sessions 7 and 8. The two methods compared were: (1) selecting each cell in a matrix using a mouse, and typing in numbers; (2) using a menu to establish an automatic cursor movement from cell to cell, then typing numbers into the cells of the matrix. The first method took longer for each cell, but the second included an overhead of a long setup time. Subjects used both methods for matrices of various sizes for one hour after their other experience with the system. Following this practice, they were asked to choose, and use, one method to enter data in each of twelve different matrices in both directions. For small matrices the mouse method was faster; for large matrices the setup time for the menu method became proportionally smaller, so that method was faster. Since subjects used each method frequently before they were asked to choose the optimal method for each matrix, it was possible to calculate the time each subject took to perform each stage of the entry methods. Consequently, it was also possible to calculate at which matrix size the temporally optimal change in methods occurred for each subject.

For all tasks performed, a timestamped videolog was preserved for all user–system dialogue and user–experimenter interaction. For the first three tests, four different analyses were then performed: (1) the time subjects took to perform the task; (2) the number of tasks achieved without major re-attempts; (3) the number of tasks achieved in a single goal; and (4) an analysis of all deviations from an optimal route for performing the tasks (for the A set, these are reported in detail in Wilson et al., 1985a). In order to make such analyses, a task analysis and model had to be constructed to specify tasks, goals, major attempts and local deviations. The explicit model used for this study contained two parts. The first assumes that users have a top-level goal to complete what is presented on the manuscript marked for editing. Within this they then have subgoals to complete each of the three tasks marked. Users then make attempts to achieve each subgoal by constructing and executing a method. The optimal method for achieving a subgoal was taken to be that which users had been shown on videotape during training; attempts include suboptimal sequences as well as this method. It is assumed in the model that at each stage in the execution of this attempt, subjects can test the consequences of their actions. If tests fail they have options: (1) to make a local correction to the attempt; (2) to construct a new attempt to complete the subgoal; (3) to postpone the present subgoal and attempt another. If all the tests in an attempt are passed then the goal is completed. This model was further specified to account for the exact nature of the local correction, re-attempts, and goal changes within the command language of the VisiOn system (this is presented in Wilson et al., 1985a, 1985b and will not be described here in detail). In general this model divided an attempt into four stages: the specification of the attempt; the establishment of the context in which to issue the commands specific to the task; the performance of commands specific to the task; and a termination of the command issuing context. Actions were associated with movement between each of the states in the model so that a local correction to an attempt consisted of returning to a higher-level menu and making another selection, using the delete key on the keyboard or similar actions which did not modify the nature of the entire attempt. The actions associated with the construction of a new attempt at the subgoal included an abort of the attempt to the top-level menu for the environment and starting the task again. There is a special case where a goal is completed by the user, but the route prescribed was not used. These were classed as 'Ignored failures'.

This formal model of the task permitted a very exact classification of the action of users which yielded results which were quantifiable. This contrasts with the informality found when observers note their reactions to a user's performance. Unsurprisingly, in the A set test the times to complete the overall tasks fell across sessions ($F(2,14) = 24.62$; $p < 0.001$) and the number of overall tasks completed without major errors increased ($F(4,28) = 3.72$; $p < 0.02$). These gross measures appear to support the assumption that general learning reduces performance time. However, a more exact analysis shows that the times to perform three tasks fell dramatically (e.g. the time to enter a formula into a spreadsheet fell from 245 to 80 seconds) while the times for the other six tasks remained constant. Similarly, a gross measure of the number of command sequences performed without major re-attempts rose significantly from 65 to 80 per cent ($F(2,14) = 4.82$; $p < 0.03$), although for five of the nine commands, and three of the eight users they did not.

The relationship here between total time and the percentage of command sequences performed without major re-attempts is not straightforward. Two of the four commands whose performance involved a decrease in major re-attempts are also associated with large improvements in time, with the other two being associated with more modest improvements. In contrast, one command sequence which shows substantial gains in the time to complete it (the sequence for typing text into a word processor) was performed with a consistently low level of accuracy throughout.

This time/accuracy variation demands a more detailed analysis of the data which shows that there was also a change in the classes of errors across sessions. The proportion of local corrections while attempting tasks doubled (see Table 1); in contrast the proportion of major re-attempts dropped by two-thirds, while the proportion of errors due to changes to goals remained constant.

Overall, there are both users and tasks for which performance remains poor, and others for which it improves. It was also true that neither the time to carry out a correct command sequence nor the number of retries after a failed command reduced significantly. These results offer little evidence for the speeding up of performance for well-formed sequences or their pro-ceduralization. Rather, the reduction in performance time is due to the reduction in major re-attempts.

It can be concluded that a view of the learning captured by this performance test which suggests that all sequences, or performance on all aspects of the system would improve synchronously, is inappropriate. Rather than a single progression, the data suggest that it would be more helpful to consider the gross performance measures as reflecting an averaging of individual command sequences each of which could lie at different stages of development within a user's repertoire of knowledge and skill. This repertoire as a whole develops, but parts of this knowledge are inaccurate, and these may persist and cause consistent errors.

Table 1. Percentages of deviations from optimal route by class

Test session	Local correction	Re-attempts	Goal changes	Ignored failures	N
A set 2	39.7	39.7	17.6	3.0	68
6	67.8	12.9	16.1	3.2	31
B set 4	37.5	40.0	15.0	7.5	40
6	64.9	27.0	8.1	0.0	37

The B set test was analysed in a similar fashion to the A set. Since the B set involved the testing of different command sequences of varying length on each session, it would be inappropriate to present time data. However, to answer the question as to whether the general improvement in system knowledge aided the acquisition of new command sequences, one can rely on error data. The number of errors for the sequences in the first B set test was constantly the same as that for the third A set test (see Table 1), whereas the error count for the A set test fell by half over the four sessions. Despite a temptation to interpret the absolute number of errors as an indication that users are recruiting chunks of knowledge established during the A set test during their acquisition of B set command sequences, this would be inappropriate since the B set sequences were not perfectly balanced for length with those in the A set. In contrast, the constancy of the error count for B set tests could suggest that the acquisition of later command sequences was not being progressively aided. However, this too would be inappropriate since the items were not the same across the B set sessions, and this would affect the relative count. However, an analysis of the pattern of errors is valid evidence, and this indicates that the learning of the later B set commands is being aided. The shift in error types across sessions for the A set test that accompanied the reduction in performance time, was from major re-attempts to local corrections. The same shift exists in the errors for the B set from sessions 4 to 6. This change in error patterns can be attributed to users learning to recognize errors in time to avoid the need for major re-attempts. This suggests, that the information in the user's repertoire which is permitting them to move from time-consuming major re-attempts on well practised commands also affects the performance of the new sequences in the B set test.

Whereas the A and B set tests were designed to indicate the development of proceduralized methods as users approach a state of expert performance, the last performance measure was intended to test how choices were made between methods when a state of expertise has been reached. The data entry task was designed to assess whether time alone was the criterion which experts used to select the best method to achieve a task as is often assumed. The method allowed the calculation of the temporally optimal entry method for any data matrix for any subject. After much practice and experience, the users then selected the method they wished to use, and used it. The choice points actually used by subjects were not the same as the temporally optimal point but showed a consistent bias in favour of the method including the use of a menu to set the direction of cursor movement over the mouse method. One explanation for this could be that users judge the relative performance times for the two methods inaccurately. However, incidental descriptions by the users suggest that they were aware of the relative performance times. This simple study of a small set of expert performance illustrates that the assumption that expert decisions are based on time alone is inadequate to explain the actual choices made. Further factors have to be introduced into models to make accurate estimates. In this study, anecdotal evidence suggests that using both a mouse to select spreadsheet cells and a keyboard to enter numbers breaks the users' concentration or is mentally effortfull. Spending time to establish the direction of cell movement on a menu and then using a keyboard to enter numbers was considered easier. Consequently, factors of cognitive load, or focus of attention must be included in models of the criteria by which experts choose methods as well as performance time if they are to be accurate.

4 TESTS OF ARTICULATABLE KNOWLEDGE

The model held of the user in this chapter incorporates a distinction between procedurally and declaratively represented knowledge. It has been assumed that procedural knowledge supports expert performance, but is not articulatable. Declarative knowledge exists in parallel with this which is articulatable. Consequently, this declarative knowledge should also be investigated as part of an investigation of the changes which take place during learning.

If users give verbal protocols during performance, one assumes that this protocol reflects both the way that users perform the present task, and the way they would perform that task outside the laboratory. Both of these assumptions have been challenged. It has been suggested that such techniques present a danger of the verbalization distorting the normal operation of cognitive processes (e.g. Ericsson and Simon, 1984). Consequently, methods of filtering protocols have been developed (e.g. Bailey and Kay, 1987) or alternatives to the simple talk aloud protocol have been suggested, either to the form and volume of data collected (e.g. Hammond et al., 1983), or in the method of eliciting protocols through tasks other than the concurrent verbalization of a single user performing a task (e.g. Suchman, 1985).

An attempt to combine these advantages in a task which probes users' knowledge was made in the prompted knowledge elicitation (PKE) test in the study of VisiOn which would give access to the users' articulatable knowledge, without imposing a concurrent protocol task. The purpose of this test was to assess the changes in verbalizable knowledge during learning, and to evaluate the relationship between this and the performance measures. In the PKE test, users are presented with pictures of screens, about which they are asked focused questions (see Canter and Brenner, 1985). The picture-based interviewing technique reduces the quantity of protocol to be analysed by presenting prespecified questions about specific sample system states. It also permits a formalization and quantification of the analysis by having prespecified sets of target claims which could be produced by users in reply to the questions. The formalization of the task and dialogue also permits a specification of the structure of the PKE task, and of its analysis. From this it is clear that users' responses are not purely expressions of their knowledge of the system, since the sources of information which are likely to influence a protocol content include the knowledge of the domain of application, information in the photograph, the interpretation of the picture probe task itself, and the prompts from the experimenter, in addition to the knowledge of the system being studied.

A detailed report of this study is presented elsewhere (Barnard et al., 1986), so only a brief outline of the method and results will be presented here. The test was administered at the end of sessions 3 and 6, where the same 16 pairs of photographs were used (data from the first nine picture pairs only are presented). Each screen was presented in two different pictures: one where domain information was present and another where it was absent. Data will be collapsed over this condition in the present description. Users were given initial instructions which put them in the position of someone explaining the system to a friend. The experimenter prompts in this study focused about: (1) the task the user would be performing to be in the state presented in the picture; (2) the command sequence that would have reached the screen portrayed in the picture; (3) the sequence from this screen that would be used to achieve the suggested task. Subjects' verbal responses were recorded and transcribed. The responses to the probes can be regarded as providing several 'statistical' samples of a user's repertoire of system knowledge at the two points in the learning programme when the test was administered. Accordingly, a scoring procedure was devised to capture the content, form and attributes of that knowledge as expressed in the verbal protocols. The scoring scheme divided the protocol into 'core propositions' or 'claims' that were identifiably different for a particular photograph. This procedure was similar to the approach adopted by Long et al. (1983) with protocol data from groups discussing computer use. Claims made by the subjects which contained information that fell within classes defined by the probes were judged as target claims, while other claims were judged to be non-target and were dropped from the analysis. Within the classes of target information, user claims were scored as: true (accurately describing system operation); false (in contrast to system operation); inexact (a claim neither clearly true nor false); or indeterminate (for ambiguous claims and explicit statements of ignorance).

The analysis of the protocols showed a rise in the mean number of target claims from session 3 to session 6 (Wilcoxon $T = 3$, $p < 0.05$) indicating that the users' verbalizable knowledge was increasing. It is possible that this rise was caused by an increased familiarity with the test, or ease in talking to the experimenter, but there are few grounds for this explanation. This overall rise is due to an increase in the number of true claims (subjects: Wilcoxon $T = 3$, $p < 0.05$; materials: Wilcoxon $T = 3.5$, $p < 0.05$), since the number of false, inexact and indeterminate claims did not rise significantly. However, within this rise there is a complex change in the exact claims which were repeated at the two sessions (46 per cent of true claims at session 6 were repeated; 31 per cent of false claims).

The rise in true claims is not unexpected, and the fact that some false claims recurred is consistent with prior observations based on protocol examples and with the repertoire view of knowledge outlined above. However, two-thirds of the false claims made in session 6 were new. Although these new claims arose for a number of reasons, more detailed analysis suggests that false claims were made in spite of known facts that were accurate. With one of the probes for example, only one user correctly identified it as the prompt state for editing cells in the spreadsheet. Although true claims were elicited concerning cells and work area actions, those concerning the task and prior menu selections nearly all focused on the entry rather than the editing of material. This confusion between menu paths was understandable in terms of their gross similarity. However, other probes elicited true claims from users which accurately described prompt states for entry. They had apparently not accessed that knowledge to block false inferences concerning the edit prompt. Thus, while the number of true claims may rise, the underlying knowledge from which the claims were derived may remain compartmentalized and functionally inaccessible in a novel context.

Having established some characteristics on the verbalizable knowledge acquired by users and changes in it during learning, it is necessary to investigate the role that that knowledge may play, and circumstances under which it may be recruited.

5 TESTS REQUIRING IMPLICIT USE OF KNOWLEDGE

There are many possible off-line recall tests, however no strong relationship between recall and the usability of computer systems can be shown (Mac-Lean et al., 1984). Despite this, if some evidence exists for a particular form of user representation being important in determining performance, other indirect measures can often provide useful converging evidence to form a more detailed conclusion. Two off-line tests were included in the VisiOn Study which were designed to elicit different aspects of users' knowledge of the system and its operation. These were a questionnaire and a command name sorting task which were administered only in session 7.

The questionnaire included 40 questions to which users had to give true/ false answers and confidence ratings on a four-point scale. Specific questions were constructed in the light of the general user performance on previous tests. The purpose of the questionnaire was to test if users would recognize states of affairs or system rules even if they could not articulate them, and conversely, to see if they would endorse false claims that were consistent with common performance errors. The mean correct response rate was 70 per cent, and of the 96 errors there were 73 false positives and 23 false negatives, with 11 questions accounting for 63 per cent of the errors.

Of the questions which were consistently answered wrongly, some support user performance errors and others contradict them. One question asserted that the terminator 'done' occurred on all command menus. This is false, and yet was affirmed by 5/8 subjects (with a mean confidence of 1.6; 1 = very sure, 4 = complete guess). The item 'done' was responsible for a large number of performance difficulties in the A set test, but the performance problems were mostly due to users forgetting to select it when available, rather than attempting to select it when it was not available. Since the statement contradicts performance, the affirmation indicates uncertainty rather than the presence of false knowledge. This affirmation of a falsehood supports the idea that where there are performance difficulties, they are due to ambiguity or uncertainty in the subject's mind. In the light of this, it is surprising that users were extremely confident in the questionnaire that all their answers were correct (mean confidence = 1.35).

The most common user error in the A set test was associated with typing text into an editor. This is the only task for which no menu selection need be made—it is the default state. However, many users searched the menu hierarchy for a command which would allow them to do this. The only question which all users answered incorrectly (with a mean confidence of 1.9), stated that there was an option at a specified point which enabled the typing of text. This again is totally consistent with performance errors; however, all users managed to type text for each of the A set tests and on many other occasions although they never made a menu selection. Further, if they had a knowledge of the menu hierarchy at the point specified in the question (which they had frequent experience of) they would have ruled out the possible presence of a menu item there. These two examples both support the view that users are willing to endorse false statements about the system which are consistent with user errors. Such endorsements of false statements provide useful information about areas where errors in performance are likely to occur.

Both the PKE task and the questionnaire suggest that users are not good at recalling the menu structure in order to rule out interpretations. The name sorting task was designed to test incidental learning of the menu structures of the system. These data would supplement those for the C set performance data, and the PKE, to support a view of the representation that could support problem-solving behaviour.

The VisiOn system contains three environments. Each environment uses a top-level menu to select commands. Many of these menu selections result in the presentation of a second menu list. The 'cue' component of the task comprised the main menu lines from each of the three environments. The list of stimulus words to be sorted under each of these headings was taken from the menus that arise when menu items on these lists are selected. A total of 72 secondary menu items were selected from four menu headings in each of the environments. Subjects were presented with a large sheet of paper on which were headings consisting of the main menu lines from the system. They were also given an alphabetic list of the 72 target words. They were instructed to write each target word under the menu item that would be selected to get that word on a menu. Out of 576 items there were 314 assigned to some heading; of these 141 were assigned correctly and 173 incorrectly. For the items that were used in the A set test, on average 6 subjects placed them correctly. For items used in the B set test, on average

3.3 subjects placed them correctly. Of the items not used, those that occurred on menus with the A set items were placed correctly by 1.6 subjects on average; those that occurred on menus with B set items were placed correctly by 0.8 subjects; and those that occurred on C set menu were placed correctly by 0.1 subjects.

This evidence suggests that users are developing some representation of the menu structure and that the accuracy of this representation corresponds to the frequency of usage of a particular menu. However, it would appear that there is very little representation of items that have not actually been used themselves. The conclusion that can be drawn from this test is that users do not represent an abstract menu hierarchy incorporating a rich representation of unused items which could be recruited to solve novel problems.

6   INTERRELATIONSHIP OF THE TESTS

This chapter has briefly described the results from six assessments of user knowledge which provided sufficient data to be analysable. There are three aspects of the interrelationships of these tests which must be considered. Firstly, how do the test results combine together? Secondly, do the results indicate important individual differences? The third aspect, concerning what the results from the tests as a whole indicate about the learning of computer systems will be discussed in the next section.

The initial view taken of user learning distinguished between verbalizable and non-verbalizable knowledge. Consequently, the PKE test was used to access verbalizable knowledge and performance tests were used to test non-verbalizable knowledge. If these tests access bodies of knowledge which are recruited at different times then their data should be incompatible. Users can accurately perform tasks in the A set, whereas they cannot either describe them in the PKE, or assign their menu items in the sorting task. This contrast supports the argument that performance and non-performance tests either access different knowledge, or access the same knowledge differently (e.g. Ericsson and Simon, 1984). The objective of this study was to investigate the representations established during learning, but for practical purposes only the representations which support performance may be of interest. Consequently, is there any evidence for the knowledge illustrated in the tests of verbalizable knowledge, or those involving an implicit use of knowledge, supporting performance?

There are cases where the evidence from different tests coincide. However, these combinations of evidence result from interpretation, since the data themselves are of different forms. For example, common errors in the A set, are associated with false confirmations in the questionnaire; similarly, commonly used functions in the performance tests are associated with correctly placed menu items in the sorting task. In both these cases, there is confirmatory evidence from the tests requiring an implicit use of system knowledge for findings from the performance tests.

One measure of the overlap between tests, is to rank the eight subjects from the 'best' to the 'worst' on each of the measures used, and compare the extent to which the relative ranks overlap. The rankings for selected measures are presented in Table 2. From this, there is a significant correlation between the PKE and the A set performance test ($r_s = 0.85$; $p < 0.01$), showing that users who displayed accurate performance also demonstrated more accurate verbalizable knowledge of the system than those whose performance was less accurate.

The highest correlation between tests is between the questionnaire and the A set test ($r_s = 0.97$; $p < 0.01$). The measure from the A set, is one of the users' ability to construct or retrieve methods which will successfully achieve the task demands, and to perform these without major errors. This appears to contrast with the questionnaire where users confirmed or denied generalizations about system function. However, the questionnaire was designed to confirm observations made during the A set test so the content of the two is linked. Whereas this test does indicate that the types of knowledge tested for in the questionnaire could be those which support performance in the A set test, it is also possible that although they test different knowledge, subjects who are good at one test are good at the other.

The measure which is the worst correlate with the other appraisals is the speed of keying in the data entry task. However, since all the other measures sample learning, and the data entry task samples expertise on a predominantly manual task, this is not surprising. It is, however, interesting to note that user 8 shows good performance on the learning tasks while his performance on both aspects of this more expert task is poor. This contrasts with subject 7 whose performance on the learning tasks was also consistently good, but whose performance on the expert task was correspondingly good.

There is little in this small population sample that can be used to argue about individual differences. The experimenters' observations of the user during the experiment suggested that user 5 had considerable difficulty with all the tasks, and also she is ranked last overall in Table 2. The observed difficulty manifested itself in an unwillingness to act without reassurance. However, her rank is very close to that for subject 6, who did not show such obvious difficulties. It is tempting to argue that there could be three populations represented in Table 2. Subjects 5 and 6 show the worst overall performance, subjects 7 and 8 show the best performance and the remaining four subjects yield average data. However, the sample is too small for such a conclusion to have more than face validity.

Table 2. Intersubject differences and variability. The A set performance test is reported by the percentage of tasks completed at the first attempt. The B set performance test, by the total number of errors made by subjects. The recall test, by the number of items correctly assigned. The questionnaire, by the number of questions correctly answered. The **PKE** test, by the number of true core claims produced. There are two measures derived from the data entry task: the time to complete simple manual tasks (a human simple speed measure) and a distance measure from the point of optimally efficient trade-off (T/O) for each subject (users 5 and 6 always chose one method, hence their trade-off point is at infinity).

Ranks and Ranges

Subjects	A set tasks at 1st attempt	B set number of errors	Recall (no.)	Quest (no.)	PKE (claims)	Data entry Completion time (s)	T/O point	Mean rank
1	5	8(61)	6	5	6	3	3	5.1
2	6	5	8(7)	6	7	4	2	5.4
3	4	4	4	3.5	4	8(83)	4	4.5
4	3	2	5	3.5	3	2	5	3.4
5	7	6	7	7	8(22)	7	7.5	6.7
6	8(51%)	7	3	8(23)	5	6	7.5	6.3
7	2	3	2	1(33)	2	1(54)	1(0.6)	1.7
8	1(98%)	1(7)	1(30)	2	1(90)	5	6	2.4

7 SUMMARY OF INDICATIONS CONCERNING USER LEARNING

The initial view taken of user learning was based on a three-stage model (after Fitts, 1964 and Anderson, 1983) of a reduction in performance time and errors with task experience arising from the chunking of knowledge (after Rosenbloom and Newell, 1987) which results in a state of expert performance where the choices between methods to achieve tasks are made on the basis of performance time (after Card et al., 1983). Should the present data motivate any changes in this view?

The data from the A set test show no evidence of a simple reduction in performance time during learning. Instead, it appears that there are some tasks and some users for which there is improvement and others for which there is not. This cannot be used to argue against the initial view taken of learning, since it may be that the users in this study did not reach a stage of proceduralization even for established tasks. In order to interpret these data it was suggested that user knowledge should be viewed as a repertoire of knowledge fragments which changes during learning, and which is sampled by the various tests.

The A and B sets sampled different stages in the development of the repertoire. The changes in the error patterns for these two command sets follow the same pattern, despite one set having been consistently practised while the other was not. This indicates that the change in error pattern was taking place over the sessions rather than over practice with the individual commands. This change was from the frequent occurrence of errors that led to the performance of new attempts to perform tasks, to the dominance of errors which could be corrected locally. From this it would appear that the dominant change in knowledge was either that which led to the application of tests during attempted execution in order to identify errors, or the ability to recover from errors locally.

Both the A set and the off-line tests illustrate the development of user knowledge on core commands as learning progresses. The PKE test also showed that as learning progresses not only is false information removed from the repertoire, and more true information acquired, but new false information is also acquired.

The data from the sorting task add to this picture of a developing repertoire by suggesting that users had not developed a representation of the menu hierarchy even by the end of the experiment, but were performing methods in a sequential manner. Although the users' knowledge of the menu hierarchy was better for frequently used commands than for those which could have been acquired incidentally, rules about the structure of the hierarchy were also being learned. For example, the questionnaire results showed that users affirmed rules that a terminator 'done' appeared on all menus, and that all methods required an initial menu selection, even though these were false generalizations. These inaccurate rules appear to be applied in many of the attempts which result in user errors in the performance tests. Consequently, it would appear that errors in performance arise when there is a conflict between elements in the repertoire, leaving an ambiguity in the creation of attempts to perform tasks. This ambiguity is resolved by the application of a high-level rule which may be inaccurate, resulting in an attempt which results in error when enacted.

From these data as a whole a picture has developed of a repertoire of knowledge which changes during learning. The repertoire contains fragments of performable methods, high-level system rules and some structures representing commonly used commands. Some of these are accurate to the system, others are not. As learning progresses, some of the method fragments may combine, but they do not do so for all methods at the same rate. The most significant change is in fragments that permit recovery from error states which become more accessible when those states are encountered. There is also a loss of false information but new false information is acquired, which may be contradictory to other knowledge fragments. When users have to perform a task by constructing an attempt and encounter a conflict they default to a system rule although it may be inaccurate.

8 CONCLUSION

No simple technique can be expected to capture adequately the complexity of using a computer system, although it may provide useful local information about a specific aspect of performance. However, a range of carefully related different measures can give sets of detailed information which provide converging evidence on which to base a more detailed conclusion. This chapter has shown the application of such an approach in one study for which it has proved to be extremely successful in increasing understanding of how various aspects of the user interface affect both user performance and the underlying mental representation on which it is based.

The representation that appears to underly the learning captured by these tests is best described as a repertoire of knowledge fragments. This repertoire develops over the course of learning, and different measures and performance tasks sample different sets of this repertoire.

ACKNOWLEDGEMENTS

The research reported was performed while the authors were employed by the MRC Applied Psychology Unit, Cambridge, and was supported by a research grant from IBM Human Factors Laboratory, Hursley, UK.

REFERENCES

Anderson, J. R. (1983). *The Architecture of Cognition*. MA, USA: Harvard University Press.

Bailey, W. A. and Kay, E. J. (1987). Structural analysis of verbal data. In: *Proceedings of CHI + GI '87*, Toronto, 5–9 April. J. M. Carroll and P. P. Tanner (eds). New York: ACM.

Barnard, P. J., Hammond, N., Morton, J., Long, J. and Clark, I. (1981). Consistency and compatibility in human–computer dialogue, *International Journal of Man–Machine Studies*, 15, 87–134.

Barnard, P. J., Wilson, M. D. and MacLean, A. (1986). The elicitation of system knowledge by picture probes. In: *Proceedings of CHI '86*. New York: ACM.

Canter, D. and Brenner, M. (1985). *Uses of the Research Interview*. London: Academic Press.

Card, S., Moran, T. and Newell, A. (1983). *The Psychology of Human–Computer Interaction*. Hillsdale, NJ, USA: Erlbaum.

Carroll, J. M. and Carrithers, C. (1984). Blocking learner error states in a training wheels system. *Human Factors*, 26(4), 377–89.

Catrambone, R. and Carroll, J. M. (1987). Learning a word processing system with training wheels and guided exploration. In: *Proceedings of CHI + GI '87*, Toronto, 5–9 April. J. M. Carroll and P. P. Tanner (eds). New York: ACM.

Ericsson, K. A. and Simon, H. A. (1984). *Protocol Analysis, Verbal Reports as Data*. Cambridge, MA, USA: MIT Press.

Fitts, P.M. (1964). Perceptual-motor skill learning. In: *Categories of Human Learning*. A. W. Melton (ed.). New York: Academic Press.

Hammond, N., MacLean, A., Hinton, G., Long, J., Barnard, P. J. and Clark, I. A. (1983). *Novice Use of Interactive Graph-Plotting System*. (Technical Report HF083). Hursley, UK: IBM Human Factors Laboratory.

Lemmons, P. A. (1983). Guided Tour of VisiOn. *Byte*, 8 (6), 256–78.

Long, J., Hammond, N., Barnard, P. J. and Morton, J. (1983). Introducing the interactive computer at work: The user's views. *Behaviour and Information Technology*, 2, 39–106.

MacLean, A., Barnard, P. J. and Hammond, N. (1984). Recall as an indicant of performance in interactive systems. In: *INTERACT '84*. B. Shackel (ed.). Amsterdam: North-Holland.

MacLean, A., Barnard, P. J. and Wilson, M. D. (1985). Evaluating the human interface of a data entry system: user choice and performance measures yield different tradeoff functions. In: *People and Computers: Designing the interface*. P. Johnson and S. Cook (eds): Cambridge, UK: Cambridge University Press.

Neisser, U., Novick, R. and Lazar, R. (1963). Searching for ten targets simultaneously. *Perceptual and Motor Skills*, 17, 427–32.

Neves, D. M. and Anderson, J. R. (1981). Knowledge compilation: mechanisms for the automatisation of cognitive skills. In: *Cognitive Skills and Their Acquisition*. J. R. Anderson (ed). Hillsdale, NJ, USA: Erlbaum.

Newell, A. and Rosenbloom, P. S. (1981). Mechanisms of skill acquisition and the law of practice. In: *Cognitive Skills and Their Acquisition.* J. R. Anderson (ed.). Hillsdale, NJ, USA: Erlbaum.

Roberts, T. L. and Moran, T. P. (1983). The evaluation of text editors: Methodology and empirical results. *Communications of the ACM,* 26, 265–283.

Rosenbloom, P. S. and Newell, A. (1987). Learning by Chunking, a production system model of practice. In: *Production System Models of Learning and Development* D. Klahr, P. Langley, R. Neches (eds). Cambridge, MA, USA: MIT Press.

Snoddy, G. S. (1926). Learning and stability. *Journal of Applied Psychology,* 10, 1–36.

Suchman, L. A. (1985). *Plans and Situated Actions.* (Research Report ISL-6). Palo Alto, CA, USA: Xerox.

Wilson, M. D., Barnard, P. J. and MacLean, A. (1985a). *User Learning of Core Command Sequences in a Menu System.* (Technical Report HF114), Hursley, UK: IBM Human Factors Laboratory, p. 117.

Wilson, M. D., Barnard, P. J. and MacLean, A. (1985b). Analysing the learning of command sequences in a menu system. In: *People and Computers: Designing the Interface.* P. Johnson and S. Cook (eds). Cambridge, UK: Cambridge University Press.

WHAT KIND OF INFORMATION DO USERS USE TO OPERATE A TEXT EDITING SYSTEM?

R. SCHINDLER and A. SCHUSTER

Department of Psychology, Humboldt University, Berlin, GDR

1 THE PROBLEM

Modern interactive computer systems can perform a wide variety of functions but can also be very complex. To tackle this complexity the user has to acquire a good deal of knowledge. Otherwise he/she may reduce the system's complexity and thus not make use of all the available functions. As a result, some of the system designers' work is rendered useless. Hence, not only novice users but also skilled users need assistance in learning a system. One prerequisite for effective assistance is to specify the knowledge that users require in learning to use an interactive system.

From experiments in which we analysed the process of self-guided user's learning task completion with the help of a text editing system (Schindler, 1987, 1989; Schindler and Schuster, in preparation) the following can safely be said in summary.

1. To build their knowledge structures about the text-editor subjects needed four classes of information:
   a. *Verbal phrases* to label the to-be-delegated object transformations (W). Such phrases consisted of a verb representing a certain transformation (e.g. delete, load, insert) and a noun representing the object that the transformation acted on (e.g. character, program, word).
   b. *Combinations of alphanumeric features* displayed on the screen, or word-markers to label them (X). They were used by subjects to represent both the initial and the anticipated perceptible state of the system.
   c. *Units of action* (HE), which represent the physical actions that the user has to perform to accomplish an anticipated object transformation. In our investigations the units of action consisted of a single keystroke or of sequences on the keyboard.

COGNITIVE ERGONOMICS:
UNDERSTANDING, LEARNING AND DESIGNING
HUMAN–COMPUTER INTERACTION

d. *System processes* (SV), which represent task-relevant processes of the system that are not made visible. They were used by subjects for understanding why a particular unit of action, a specific transformation or state of the system had to be realized in the completion of the text editing tasks.

These four classes of information represent the task-relevant entities of the system that subjects required to build up their knowledge structures about the text editing system. According to Tauber (1986), they can be regarded as the system's components that the user's virtual machine consisted of. We call them *elements of interaction*.

2. The experiments performed made clear that the subjects acquired their knowledge structures with regard to the single tasks that they had to complete in reaching the top-level goal (text editing), i.e. they structured the items of information to be learnt in terms of the mental operations and physical activities that they had to perform while completing the subtasks that the text editing tasks consisted of. Both of the most advanced approaches to modelling the user's knowledge—the Cognitive Complexity Theory (Kieras and Polson, 1985), and the Task-Action Grammar (Green et al., 1987)—are also based on a task-oriented knowledge organization. Klix (1985, 1986) called such knowledge situational or event-related knowledge.

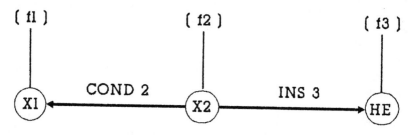

```
IF      X1: f1
THEN    X2: f2
        HE: f3
```

Figure 1.   A unit of interaction represented by a semantic network and by a condition-action rule. f1, f2, f3 denote the properties of the elements of interaction. The labelled arcs indicate the role (functional meaning) of the respective elements of interaction in the unit of interaction. COND: condition; INS: instrument; X1: current system state; X2: system state displayed on screen if the unit of action (HE) is correctly performed.

Regarding human–computer interaction, we named these conceptual entities *units of interaction*. In Figure 1 an example is given to illustrate how we can represent units of interaction. As the role that the elements of interaction play in completing a dialogue step is made more clear by a semantic network representation we will use graphical representation here.

In the course of learning, the units of interaction are changed through two processes. The first process, generalization, is based on recognized similarities between different units of interaction and creates a new unit, which captures what the individual units of interaction have in common (Anderson, 1982). As a result of the second process, composition, the number of units of interaction can be reduced by building 'macro-units'. The creation of such macro-units is based on recognized regularities in sequences of transformations.

3. To understand how the single tasks were to be completed with the help of the text editing system, the subjects made very flexible use of the four elements of interaction, i.e. they combined them flexibly to different types of units of interaction.

The present investigation is based on these experimental findings. We are interested in a more accurate analysis of the meaning that the four elements of interaction have in promoting the user's understanding of how to operate the text-editor. Intuitively we might suspect that the more detailed the information that the user is given about the system the better might be the user's understanding; the present investigation focuses on this.

2 METHOD

Three groups of subjects were used, each of which consisted of six undergraduate students of psychology with no relevant knowledge about the domain but with similar verbal intelligence and capability in analogous reasoning (SASKA test; Riegel, 1967). The experimental procedure consisted of four phases.

Phase one

In the first phase, subjects were instructed on the completion of a complex text editing task (41 dialogue steps had to be performed to complete it correctly) with the help of three different types of units of interaction.

a. One group (B1) was taught how to complete each of the 41 dialogue steps with the help of units of interaction that consisted of only three elements of interaction: a combination of alphanumeric features that characterized the current state of the system (X1), a unit of action (i.e. a single keystroke or a sequence of keystrokes on the keyboard that was conditionally related to the system state), and a second combination of alphanumeric features displayed on the screen if the unit of action was correctly performed (X2; see Figure 2).

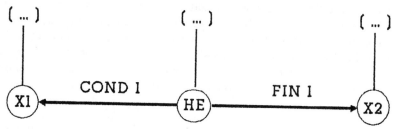

Figure 2.   The type of the units of interaction used to instruct the B1 group. The brackets represent the properties of the elements of interaction. X1: current system state; X2: anticipated system state; COND: condition; FIN: finality.

b. The second group (B2) was in addition given verbal phrases representing the kind of transformation to be realized in each dialogue step. Figure 3 shows that this type of unit of interaction gives a more precise explanation of the dialogue steps to be performed. Both the verbal phrases representing the transformation (W) and the system states (X2) can play the role of the user's actual goal (indicated by FIN 1 and FIN 2). Also, the system state (X2) can be considered as giving a deeper explanation of the transformation to be performed (represented by ATT 1).

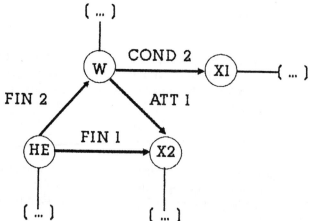

Figure 3.   The type of the units of interaction used to instruct the B2 group.

c. The third group (B3) was instructed on the completion of each dialogue step with the help of the most complex unit of interaction. They were also taught the system process (SV) going on within the machine if the unit of action (HE) was correctly performed (see Figure 4).

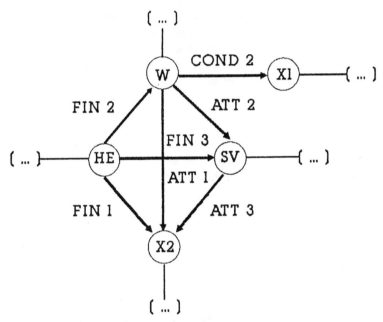

Figure 4.   The type of the units of interaction used to instruct the B3 group. INS: instrument.

Phase two
After this phase of instruction the subjects from all three groups had to repeat the first text editing task until it was correctly performed. In this phase, any lack of information could be supplied by asking the experimenter.

Phase three
In the third experimental phase, the three groups had to learn the completion of three other text editing tasks through asking the experimenter questions. The order of the three tasks was decided by the experimenter and was the same for all the subjects. The subtasks that these tasks consisted of were in parts either completely different, or similar, or identical to that of the instructed one. To achieve precise control of the information searching and using behaviour the subjects were not allowed to operate the system directly. They had to verbalize everything they wanted to know and intended to do.

The experimenter answered the subjects' questions and provided feedback about the correctness of the subjects' performances. It is clear that this experimental procedure did not simulate learning by exploration. The dialogue between subjects and experimenter was recorded on tape. Each of the three learning tasks had to be repeated until each subject met a predefined learning criterion: 80 per cent of the subtasks that the tasks consisted of had to be completed correctly and without any help from the experimenter. To focus the subjects' attention on the understanding of the task completion with the help of the device, the experimenter informed them that after the learning experiment they had to explain to an uninitiated person how to operate the system.

Phase four
In the last experimental phase, the subjects from the three groups had to complete a post-test. This consisted of subtasks that subjects had already learnt in one of the former experimental phases.

3   RESULTS

The results obtained are presented in accordance with the experimental phases realized.

3.1   Task Completion in the Second Experimental Phase

We were interested in two questions:

1. Did the elements of interaction that the instructed units of interaction consisted of guide the subjects' information using and searching behaviour in the second experimental phase?
2. To what extent were the subjects in the three groups able to repeat the completion of the first editing task correctly?

To answer the first question we analysed the degree to which the subjects in the three groups made use of (i.e. reproduced, produced, or asked for) the elements of interaction in completing the first editing task. It is clear that at all events the subjects had to realize the units of action to complete a dialogue step correctly. Hence, we only analysed the use of the other three elements of interaction.

Figure 5 shows that the groups B2 and B3 used verbal phrases representing the object transformation to be delegated (W) to a greater extent than B1 group did. Furthermore, group B3 used system processes (SV) to a larger

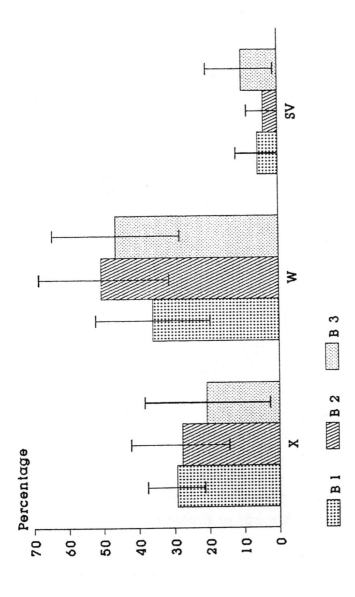

Figure 5. Elements of interaction used by subjects to complete the first editing task. X: system state; W: verbal phrases representing the to-be-delegated transformation; SV: system processes.

degree than the other two groups of subjects did. It can also be seen in Figure 5 that in the second experimental phase all three groups also used (i.e. produced or asked for) elements of interaction that were not given in the phase of instruction.

To answer the second question, we calculated the rate of assistance needed by subjects in task completion. The rate of assistance is the sum of the rate of errors and the rate of information questions that subjects asked while completing the dialogue step. It represents the degree of external help that subjects needed in completing a task.

It can be seen in Figure 6 that the rate of external assistance needed by the three groups tended to depend in a negative monotonous relationship on the complexity of the units of interaction taught in the first experimental phase. However, only the difference between the B1 and B3 group is significant.

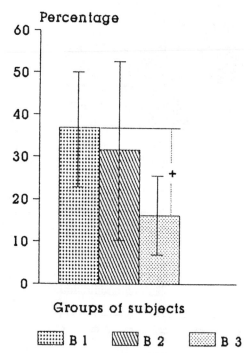

Figure 6.   Rate of external assistance needed to complete the first editing task.

3.2 Phase Three

Here we were interested in the effects of the units of interaction instructed in the first experimental phase on the subsequent phase of self-teaching of

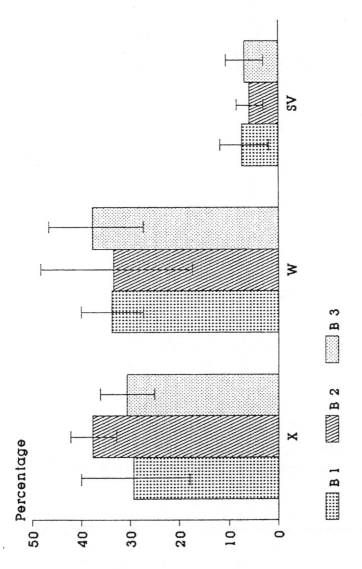

Figure 7. Elements of interaction used in the phase of self-teaching.

the three other text editing tasks. As previously, we first analysed the information using and searching behaviour of the three groups of subjects. Figure 7 shows the extent to which the experimental groups made use of (i.e. produced, reproduced, or asked for) the elements of interaction in building their knowledge structures. There were no significant differences between the three groups of subjects. In the phase of self-teaching the subjects' information using and searching behaviour did not depend on the units of interaction taught in the first experimental phase.

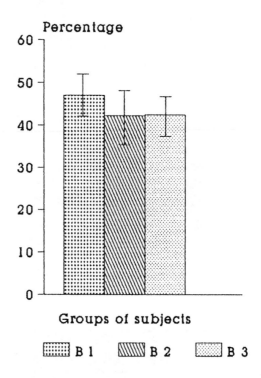

Figure 8.   Rate of external assistance needed by the groups in their first attempts of task completion.

In their first attempts to master the three tasks the B1 group tended to need more external assistance than the other two groups (see Figure 8); however, the difference was not significant. In their second attempts to complete the given tasks there were significant differences between the three experimental groups. The B1 group needed more external assistance than the two other groups, whereas there was no significant difference between groups B2 and B3 (see Figure 9).

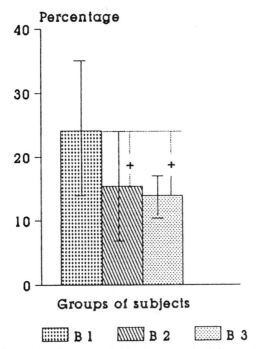

Figure 9.   Rate of external assistance needed by the groups in their second attempts of task completion.

3.3 Phase Four

The last result we are going to present here concerns the rate of external assistance needed by the three experimental groups to complete the post-test. Figure 10 shows that there were no differences between groups B2 and B3; group B1, however, needed more external assistance than the other two.

4 DISCUSSION

Only at the very beginning of their learning was the subjects' information searching and using behaviour influenced by the kind of instructions given in the first experimental phase (see Figure 3). In the course of learning the subjects from all three groups required exactly the same elements of interaction to build their knowledge structures about the text-editor (see Figure 6). This result corresponds well with other experimental findings regarding the influence of prior knowledge on individual learning (Waern, 1986).

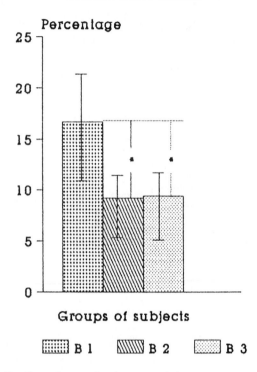

Figure 10.   Rate of external assistance needed to complete the post-test.

Regarding a user's understanding of task completion with the help of the system, the four elements of interaction have a different meaning. Subjects made use of system processes in a very small number of dialogue steps. This holds true also for the B3 group which was taught this kind of information in the first experimental phase (see Figure 6). Thus, system processes cannot generally be regarded as task relevant. As illustrated in Figure 4, system processes can have different functional meanings. They can either play the role of the user's goal (indicated in Figure 4 by the finality relation FIN 3) or/and give a more detailed explanation of the transformation to be performed (indicated in Figure 4 by the two attribute relations ATT 2 and ATT 3). It is evident from the experiments that only if there were inconsistencies in the human–computer interface, would the subjects make use of system processes to deepen their understanding of the to-be-delegated transformation. In general, subjects represented their actual goal by verbal phrases for the kind of object transformation (W) required in completing a dialogue step. The subjects only made use of system processes to represent their actual goal if the transformation was not an element of the external task space; for instance, if the transformation to be delegated to the system was not directly aimed at changing a task object (e.g. a subprogram had to be loaded or finished).

In all the experimental phases realized, B1 group needed the most external assistance (see Figures 6, 8, 9 and 10). This group was instructed on the completion of the first editing task with the help of units of interaction consisting of the units of action (HE) and the combinations of alphanumeric features (X1 and X2; see Figure 2). We can conclude from these results that a verbal representation of the kind of transformation to be delegated to the system is one of the most important points in the user's understanding of task completion with the help of the text-editor. Although, in the phase of self-teaching (phase three), the subjects from the B1 group used exactly the same elements of interaction as the other two groups did, they were not able to compensate their informational disadvantages from the first phase in the course of the experiment.

5 CONCLUSIONS

In order to determine the task-relevant entities of a system we have to analyse the transformations the user has to delegate to the device. Only those entities that have a functional meaning in task completion are task-relevant and enter the user's internal representation about the system. Regarding the user's understanding of task completion with the help of a device, the most important elements of interaction are: verbal phrases representing the transformations to be delegated to the system, sets of properties to identify and classify the states of the system, and the physical activities to be performed. System processes are task-relevant only if there are inconsistencies in the human–computer interface or if the transformation to be delegated is not an element of the external task space.

REFERENCES

Anderson, J. R. (1982). Acquisition of cognitive skills. *Psychological Review*, 89, 369–406.

Flammer, A. (1981). Toward a theory of question asking. *Psychological Research*, 43, 407–20.

Green, T., Schiele, F. and Payne, S. (1987). Formalisable models of user knowledge in human–computer interaction. In: *Theory and Outcome in Human–Computer Interaction*. T. Green, J. Hoc, D. Murray and G. van der Veer (eds). London: Academic Press.

Kieras, D. E. and Polson, P. (1985). An approach to the formal analysis of user complexity. *International Journal of Man–Machine Studies*, 22, 365–94.

Klix, F. (1985). Über die Nachbildung von Denkanforderungen, die Wahrnehmungseigenschaften, Gedächtnisstruktur und Entscheidungsoperationen einschliessen. *Zeitschrift für Psychologie mit Zeitschrift für Angewandte Psychologie*, 193 (3), 175–211.

Klix, F. (1986). Memory research and knowledge engineering. In: *MACINTER I*. F. Klix and H. Wandke, (eds). Amsterdam: North-Holland.

Riegel, K. F. (1967). Der sprachliche Leistungstest SASKA. Goettingen.

Schindler, R. (1987). Prior knowledge and the efficiency of self-guided user's learning. *Zeitschrift für Psychologie mit Zeitschrift für Angewandte Psychologie*, suppl. 9, 74–81.

Schindler, R. (1989). Wissenserwerb und-nutzung in der Mensch-Rechner-Interaktion: Experimentelle Untersuchungen zur Gestaltung von Benutzerschulungen. *Zeitschrift für Psychologie mit Zeitschrift für Angewandte Psychologie*, 197, 351–385.

Schindler, R. and Schuster, A. (1987). On the relationship between a user's self-teaching and his knowledge. *Sixth Interdisciplinary Workshop on Informatics and Psychology*, Schärding.

Tauber, M. (1986). Top-down design of human–computer interfaces. In: *Visual Languages*. S. Chang, T. Ichikawa and R. P. A. Ligomenides. New York: Plenum Press.

Waern, Y. (1986). On the role of mental models in instruction of novice users of a word processing system. HUFACIT report, No. 6. Department of Psychology, University Stockholm.

UNDERSTANDING THE COGNITIVE DIFFICULTIES OF NOVICE PROGRAMMERS: A DIDACTIC APPROACH

Renan SAMURÇAY

CNRS-Université de Paris 8, URA 1297 'Psychologie Cognitive du Traitement de l'Information Symbolique', Equipe de Psychologie Cognitive Ergonomique, 2, Rue de la Liberté, F-93526 Saint-Denis, Cedex 2

1 INTRODUCTION

This chapter will focus on one specific aspect of human–computer interaction: learning to program. There are two major justifications for cognitive analysis of the programming learning process. First of all, there is clear practical value in analysing programmer productivity for economic reasons and because the teaching of programming is now an integral part of school curriculums. Secondly, from a theoretical point of view programming as an activity is in itself a potentially rich source of information for cognitive psychology because programming is a specific problem-solving domain as well as being a conceptual field in its own right, with a recognized set of concepts and procedures.

This chapter will pay attention on the conceptual aspects of programming, and explore the initial stages of the learning process in novices of a number of concepts in programming. The acquisition of some aspects of the conceptual field of iteration will be examined. These aspects are concerned in particular with repetitive and iterative coding of simple procedures. The learning process will be examined through a study of the classroom teaching context exploiting problems used during the course of instruction.

This didactic approach to knowledge acquisition is designed to identify the cognitive difficulties learners encounter, and define those features of the teaching situation which are conducive to helping learners overcome these difficulties.

The theoretical framework underlying the experimental study is presented first, followed by an analysis of iteration as a conceptual field, i.e. how

COGNITIVE ERGONOMICS:
UNDERSTANDING, LEARNING AND DESIGNING
HUMAN–COMPUTER INTERACTION

iteration enters into problem-solving, and the relationships between iteration and related concepts. The final sections deal with data analysis and a discussion of the implications of these findings.

2   LEARNING TO PROGRAM AS A SPECIFIC KNOWLEDGE DOMAIN

Investigation of any learning process requires a referential system of knowledge. Programming, like other problem-solving activities in fields such as physics or mathematics, can be analysed in terms of expert performance in the domain (expert-oriented framework) or in terms of the explicitly constituted knowledge in the domain if one exists (science-oriented framework). Knowledge of this latter type is socially and historically constructed, and serves to define a system of normative and symbolic representations used for communicative purposes among individuals in the field. Mathematics is one example of a field where there is both an expert-oriented framework and a science-oriented framework. In other professional domains, however, such as industrial process control, there is no explicit knowledge base independent of the individuals involved in the domain, and expert knowledge must be solicited.

Analysis of the activity of programming requires integrating both these frameworks. Exclusive reference to experts or to explicit knowledge cannot provide access to programming in its full complexity and diversity.

2.1   What Makes the Activity of Programming Specific?

The specificities of programming as a problem-solving situation can be summarized as follows. The goal of programming is to find a solution, as in classical problem-solving situations, but also to make explicit the procedure leading to the solution. In addition, programming is mediated by a technological tool which requires the construction of functional representations on the part of the user; the functional representation is defined here as opposite to structural representation. These representations involve the elaboration and expression of the procedure in a specific 'action code system' that necessitates the acquisition and use of specific programming concepts.

These features of programming as an activity have considerable repercussions for the learning process. There is little pedagogical value in examining the process of learning to program from the point of view of the acquisition of syntax or methods. Research on the acquisition of programming methodology in novices (Hoc, 1981, 1983) has shown that the absence of a Representation and Processing System (RPS) (Hoc, 1988) intimately related to computer operation is an obstacle to their learning and use of methods which have highly specific procedural constraints. This type of result suggests that one way of designing the initial stages of the learning process is to give learners the opportunity to develop an RPS organized as a plan hierarchy. However, the acquisition of plans is in itself tied to the acquisition of programming concepts such as variable, iterative and conditional structures.

2.2   What are the Specificities of Learning to Program?

Recent studies have shown that learning to program, even on an extremely simple level, is difficult. One way of defining this difficulty is analysis of bugs in novices' programs as indicators of representations and learning problems. This generates the related issues of the identification and interpretation of bugs. The same bug can be interpreted in different ways depending on the theoretical and methodological framework used to analyse it.

Bug research has produced two categories of formalizations. The first includes theories which make no reference to the psychological mechanisms underlying programming activity. For instance early descriptive theories were based on the syntactic and semantic categories of bugs (Youngs, 1974). The second covers theories which do not integrate the specificities of programming knowledge and examine general psychological mechanisms such as interpreting bugs as arising from limitations on working memory (Anderson and Jeffries, 1985). Recently studies have attempted to combine these two perspectives with the aim of developing a more integrated approach to specific knowledge in programmers (Soloway et al., 1982; Bonar and Soloway, 1985). According to these authors, bugs occur when a single plan is used to achieve multiple goals. They define a plan as a method to accomplish a particular goal. When the goal is, for example, to count the number of occurrences of an event, the plan can be a counter-variable (i.e. initialization, updating and, if necessary, testing of a variable defined as a counter-variable).

The richness of this approach stems from the fact that it takes representations related to both the problem domain and the programming domain into account, and that multiple explanations each related to the specific context in which bugs appear are put forward.

Knowledge about the bugs that novices produce is crucial to an understanding of how their actual programming knowledge operates during problem solving (Spohrer et al., 1985). However, it is difficult to use these data (reflecting one point in novices' knowledge) to make inferences about the evolution of representations during the learning process and by extension to use this information to improve teaching methods. What are the relationships between bugs, and the teaching situations students are exposed to? Are there different levels of complexity across different types of tasks? Analyses of novices' behaviour during the learning process may help shed light on the cognitive difficulties specific to programming, and provide guidelines for teaching.

3   ITERATION AS A CONCEPTUAL FIELD

What is meant by a conceptual field? A given situation does not involve a single concept or even all the properties of a single concept. A conceptual field is a problem space whose processing entails different types of tightly connected concepts and procedures (Vergnaud, 1982). It is difficult to analyse a problem-situation involving the notion of iteration without referring to the concept of variable. The notion of conceptual field is useful in defining a meaningful class of problems for a given concept or features of a concept.

Iterative flow control, along with conditional flow control is one of the structures which allows the programmer to control program execution on logical and computational levels. Iterative structuration will be defined here globally as a response to problems whose solution requires the execution of identical actions/rules a certain number of times. The construction of an iterative plan involves the identification of the elementary actions/rules which must be repeated, and the condition governing end or continuation of the repetition.

There are two types of iterative problems which differ for the type of action planning the subject implements. These action plans do not appeal to the same cognitive activities because of the nature of the control structure involved in the problem. In the first class of problems the end-control is a constant, i.e. the number of repetitions is known in advance. In the second class, the end control is a value of a variable calculated in the loop.

Difficulties novices encounter in the construction of iterative programs vary as a function of the semantic content of the actions and/or the nature of the condition. It is likely that when the number of repetitions is known in advance, the situation is conceptually easier than situations where this number is a problem variable.

Pascal language has three forms of loop-plans: for-; repeat-; and while-. Formally these three are equivalent, since each can be written in terms of the others. The issue is thus whether they are also equivalent for the novice programmer with respect to cognitive difficulty and acquisition.

Comparison of the repeat- and the while-loops suggests that plans are not equally easy: the repeat-loop requires an action plan in the form of:

process variable/test variable

which is more like familiar procedures than is the

test variable/process variable

found in the while-loop. Subjects encounter difficulties in representing and expressing a condition about an object which they have yet to operate on. This implies that the conceptualization of the while-loop will be more difficult than the repeat-loop.

When solving loop problems, the subject must also take the variables and the operations performed on them into consideration. In terms of programming, a variable is defined as an address, but this formulation is not sufficient for analysis of the functional meaning subjects assign to this concept (Samurçay, 1989). When the values of a variable change, its labelling and relationships to other program components remain invariant. In programming with procedural languages, such as Pascal and Basic, this property is particularly apparent in looping problems having variables associated with a particular status such as

$$\text{sum} \longleftarrow \text{sum} + \text{number}$$

This type of variable will be termed 'accumulator' here to differentiate it from other types. The understanding and construction of this kind of symbolic representation requires defining this variable temporally as 'the value of the variable *sum* at $n + 1$th step is equal to its value at the nth step plus the value of the variable *number*'. The learner must designate both the preceding value and the current value (which is a value of the former) by the same name, and must process the assignment sign as an asymmetric relation. The mathematical model for variable or the equal sign are initial but insufficient as models for novices to use to implement on programming variables when these are involved in loop constructions. The mathematical description of a variable is a static one, and designates either a generic name for a set or an unknown in an equation:

$$3x = x + 10$$

Programming variables are new concepts for most students. It is likely that operating on variables in loop problems, in particular the update operation, will be difficult for them to perform successfully.

Two major hypotheses will be put forward here with regard to instructional situations. The first is that at the beginning of their learning process, students will use familiar procedures of execution developed for problem solving. Secondly, construction of procedures compatible with the operations of the system will require exposing students to problem situations where these familiar procedures do not work.

4 METHODOLOGICAL ISSUES IN EXPERIMENTAL DESIGN

The design of teaching situations lending themselves to analysis of novices' handling of loop problems presents a number of methodological options. Data on students during the learning process must be obtained. Secondly, the content of the teaching situation must be controlled as much as is feasible. One way of doing this is to construct experimental teaching situations in academic contexts where each session (class) is prepared in advance and the teacher knows beforehand what and how to teach.

In the experiment described below, two types of teaching situations were used. In the first, students were asked to solve problems involving knowledge acquired in previous sessions. In the second, students were tested on problems without having been exposed to all the concepts and procedures necessary to solve them, and the new concepts and procedures were taught as solutions to the problems.

Three groups of eighteen high-school students (aged 16–17 years) were used as subjects. They were given 25 hours of training (1 h per week) in the Pascal programming language on six SIL'Z microcomputers.

The tasks consisted of writing programs for simple arithmetic sum problems. The solutions to these problems use few repetitive actions, and are illustrative of many basic programming concepts such as loop and variable. The choice of these concepts was based on the idea that iteration is one of the most important functions in computing, and that the mastery of loop strategies is essential to the solving of crucial programming problems.

Three types of data were collected: (1) clinical observations of students' work during the problem-solving session; (2) notes taken by students; (3) structured questionnaires administered at different stages of training.

5 EXPERIMENTAL DESIGN AND OBSERVATION IN CLASSROOM SETTINGS

A series of experimental teaching situations were designed, all of which focused on problems requiring the use of iterative loops. Each situation will be described briefly, followed by a summary of the results which have appeared elsewhere, except for the last session which will receive more detailed treatment here. Data analysis was centred on a description of students' strategies in iterative procedure design.

The first session was used to identify the 'spontaneous' procedures students employ when solving a loop problem. Students were asked to write a 'natural language' procedure to calculate $x*y$ to be executed by another student on a theoretical device having only two operations: + a number, and + 1. Data analysis indicates that most students wrote loop-plans in which the order of operations was 'description of actions/repeat mark/end control'. Explicitation of the end control was not spontaneous (Samurçay and Rouchier, 1985). This may have been because spontaneous plans are associated with a 'go to' schema which is not good programming practice in Pascal.

In the second session, the repeat-loop was introduced as a way of solving problems where the number of repetitions is known, and control depends on this value. Students were asked for example to write a procedure to calculate and print out the average of 40 integers read in.

In the third session the for-loop was introduced as an economic way of performing the repeat-loop plan when the number of repetitions is known.

The tasks administered after these three sessions were designed to test generalizability of the acquired 'process variable/test variable' schema to problems where the number of repetitions is not known in advance, and the end control depends on a calculated value of a particular variable. Students were asked for example to write a procedure which read in integers until the sum was greater than 10 000, then printed out the average and the number of integers read in.

The data could be classified into four types of hierarchical strategies, each characterized by a certain level of conceptualization of variables and loop invariants (Samurçay, 1985). The first category was characterized by a lack of use of variables: the procedures were governed by results for specific data. The second category was characterized by the transcription of the word problem into an algebraic expression which did not yet contain procedural cues. The third category covered explicit procedures, with problems in formulation of programming code. The statements appeared in merged code composed of algebraic expressions, natural language constructs, and some programming code. The fourth category was characterized by the elaboration and expression of procedures involving an analysis and use of the signs of elementary actions with structured elements. The students encountered difficulties in expressing initialization and accumulation variables.

The students were given a final set of situations in which the 'test variable/ process variable' schema (the while-loop plan) was introduced as the solution to problems in which the end control depends on a particular value of a variable which is calculated during loop execution, i.e. the variable is processed if its value satisfies a given condition. The first problem was to write a procedure to calculate the average of a set of integers read in. The end of the set was indicated by the number 9999 (not to be included in the sum). The canonical solution can be obtained by using a while-loop such as:

```
                    sum := 0
                    READ(x)
                    WHILE x = 9999 DO — — — — —┐
                      BEGIN                      │
process/read┌— — — sum := sum + x              │      test/action
            └— — — READ(x)                      │
                      END            — — — — —┘
                    WRITE (sum)
```

It was predicted that subjects would use a repeat-loop and some additional instructions because they had not yet learned the while-loop plan. This was expected to yield one of two strategies:

— a strategy based on familiar procedures, characterized by a 'go to' schema (type 1). Some students, in particular those with prior experience in Basic, were expected to construct a program similar to:

```
    1   read new-value
        if new-value = 9999 then end
        else calculate sum
        go to 1
```

— an adaptive strategy (type 2). Certain students were expected to try to solve the problem by adapting the repeat-loop schema acquired previously.

Data on ten programs were obtained. Two students produced programs with the type 1 strategy, and eight the type 2 strategy. This implies that the majority of students tried to use an acquired schema. These eight programs were analysed in greater detail to identify the specific subplans used. Five of the eight programs used the read/process plan when structuring the loop body (see example in Figure 1) whereas the other three programs were based on a process/read plan that is more appropriate for while-loop plans (see example in Figure 2), as described by Soloway et al. (1982). These findings confirm earlier results obtained by the same authors on beginners' preference for a 'read/process' plan, which is more compatible with their familiar RPS.

```
REPEAT
    READ(new-value)
    sum := sum + new-value
UNTIL new-value = 9999
sum := sum - 9999
WRITE(sum)
```

Figure 1. Observed program fragment based on the read/process plan.

```
sum := 0
READ (x)
IF x = 9999 THEN WRITE(0)
ELSE REPEAT
        BEGIN
        sum := sum + x
        READ(x)
        IF x = 9999 THEN WRITE (sum)
        END
        UNTIL x = 9999
```

Figure 2. Observed program fragment based on the process/read plan.

Most students, however, first added 9999 to the sum in the loop body and then subtracted it from the sum before calculating the average, even though they apparently were using a plan similar to the while-loop. A new subsidiary problem arises when students realize this 'adaptative' solution will not work. They must then calculate the sum of the inverse of a set of integers; the end of the set is indicated by 0. Hence, if the student uses the procedure 'add first and then subtract', the solution will not work because 1/0 has not been defined. The while-loop can be introduced at this point as a satisfactory solution.

Overall, the observation data on the sessions as a whole shows that subjects' main difficulty lies in the elaboration and expression of the loop variant. Instead of a representational system in terms of transformations where the subject's activities are seen as the invariants, students must construct a representational system in terms of successive states where the invariants correspond to relationships between these states.

These observational findings were used to construct a questionnaire to test hypotheses on the clinical data (Sumurçay, 1989). The first hypothesis is related to the concept of variable. Two types of variables were defined according to the functional meaning subjects assign to them:

— *external variables* corresponding to the values controlled by the program user, i.e. values which are explicit inputs and outputs of the problem.
— *internal variables* corresponding to the values controlled by the programmers themselves, i.e. variables which are only necessary for programming the solution to the problem.

It was predicted that internal variables would be conceptually more difficult to construct than external ones. This is because internal variables require representations of computer operation in terms of system states. The second hypothesis concerns operations on variables in the loop plan: update, test, and initialization. It was expected that these operations would vary in complexity for the student. Initialization (make a hypothesis on the initial state of the system) was expected to be more difficult than updating and test operations because it is not a relevant operation in manual execution of the problem. In addition, these operations may be more or less complex with respect to the type of variables they operate upon. The updating of a counter-variable is easier than the updating of an accumulating variable. This is due in part to the fact that updating a counter-variable involves adding a constant

$$counter \longleftarrow counter + 1$$

although updating an accumulating variable requires adding another variable to the previous one.

The questionnaire was made up of a set of computer program texts where one or more instructions were missing. Missing instructions were selected on the basis of the internal/external hypotheses described above. The subjects were asked to complete the instructions. The results are summarized below:

Operations	Variable type	
	Internal	External
Initialization	1	9
Updating	12	22
Test	10	20

$N = 26$

These findings confirm these hypotheses and suggest that the subjects in the sample acquired and used counter-variable plan better than the other plans.

6 CONCLUDING REMARKS

At first glance the population studied here and the type of problem do not seem directly relevant to studies on human–computer interaction in professional contexts with more experienced subjects. However, the problems of training and learning to use complex computerized systems are important today, and a pedagogical approach which consists of designing experimental learning situations is a beneficial way of obtaining interpretable experimental data on the *process* of learning, and elaborating instructional hypotheses.

The results reported here concern the plans used by novices to solve elementary loop problems, and the acquisition features of these problems. An attempt has been made here to clarify the relationships between subjects' behaviour and the properties of the problem-situation in which they were observed.

The strategies identified here concerning novices' behaviour in programming are in line with other studies indicating that novices' initial representations are characterized by the transfer of familiar procedures in the problem domain, and mental execution of these processes. Thus, novices' program representations are a dynamic model of computer functioning. In contrast, students must develop a more static model in which the variables are represented in terms of relations between invariants, and not in terms of successive values of variables obtained through the execution process. The present results suggest that although experimental training situations were valuable for the students in the acquisition of different loop-plans, they were not sufficient for the construction of a static model. A new avenue of research in psychological studies on learning to program would be the investigation of this feature, by defining those conditions which help students to acquire programming methodologies emphasizing the analysis of the problem in terms of appropriate descriptions, rather than a description of a sequence of actions.

7 REFERENCES

Anderson, J. R. and Jeffries, R. (1985). Novice LISP errors: undetected losses of information from working memory. *Human–Computer Interaction*, 1, (2), 107–32.

Bonar, J. and Soloway, E. (1985). Preprogramming knowledge: a major source of misconceptions in novice programmers. *Human–Computer Interaction*, 1 (2), 133–62.

Hoc, J. M. (1981). Planning and direction of problem-solving in structured programming: an empirical comparison between two methods. *International Journal of Man–Machine Studies*, 15, 363–83.

Hoc, J. M. (1983). Analysis of beginner's problem-solving strategies in programming. In: *The Psychology of Computer Use*. T. R. G. Green, S. J. Payne and G. van der Veer (eds). London: Academic Press.

Samurçay, R. (1985). Programming learning: an analysis of looping strategies used by beginner students. *For the Learning of Mathematics*, 5 (1), 37–43.

Samurçay, R. (1989). The concept of variable in programming: Its meaning and use in problem solving. In: *Studying the Novice Programmer*. E. Soloway and J. C. Spohrer (eds). Hillsdale NJ: LEA, pp. 161–178.

Samurçay, R. and Rouchier, A. (1985). De 'faire' à 'faire-faire': La planification de l'action dans une situation de programmation. *Enfance*, 2–3, 241–54.

Spohrer, J. C., Soloway, E. and Pope, E. (1985) A goal/plan analysis of buggy Pascal programs. *Human–Computer Interaction*, 1(2), 163–207.

Soloway, E., Ehrlich, K., Bonar, J. and Greenspan, J. (1982). What do novices know about programming? In: *Directions in Human–Computer Interaction* B. Schneiderman and A. Badre (eds). Norwood, NJ, USA: Ablex, pp. 87–122.

Vergnaud, G. (1982). Cognitive and development psychology and research in mathematics education: some theoretical and methodological issues. *For the Learning of Mathematics*, 3, 31–41.

Youngs, E. A. (1974). Human errors in programming. *International Journal of Man–Machine Studies*, 6, 361–76.

SECTION 3

PLANNING AND UNDERSTANDING

PLANNING AND UNDERSTANDING: AN INTRODUCTION

Jean-Michel HOC

CNRS–Université de Paris 8, URA 1297 'Psychologie
Cognitive du Traitement de l'Information Symbolique',
Equipe de Psychologie Cognitive Ergonomique, 2, Rue de la
Liberté, F-93526 Saint-Denis, Cedex 2

1 INTRODUCTION

In the past, research in human–computer interaction was a well-defined topic, mainly concerning routine data acquisition tasks and creative programming tasks. There was, strictly speaking, no actual 'interaction' in these tasks, the first being entirely under computer control, the others under human control. Nowadays, computerized tasks are wider and cover almost every possible cognitive task, where the allocation of function between human and computer (or software) actually leads to interaction. Moreover, the human–computer interaction research domain is no longer as well delimited. Today, questions raised in this domain tap all cognitive psychology findings and approaches.

More and more computer systems are designed to aid professional problem-solving activities: diagnosis, programming, text production, architectural design, project scheduling, financial management, etc. The use of computers is not restricted to the definition of new routine clerical work as was the case in the past, but extends to a large variety of situations, from genuine problem-solving to the purely routine. Besides, increasingly, computer use is only part of a task which, viewed more widely, goes beyond strict human–computer interaction (Rasmussen, 1987). This is especially true in process control situations, where plant control is a main goal to be shared between human and computer (see chapter by Goodstein in this Section).

However, software designers very often consider the user as performing routine activities, even in problem-solving situations. This conception leads the user to well-known adverse outcomes, from poor solutions to outright rejection of the software. To go beyond these shortcomings, a better understanding of human expertise is needed. At the same time, the study of the interaction between expert-human and expert-computer is a good way to trace cognitive processes. From a psychological point of view, software design incorporates an implicit hypothesis of what will best serve human needs. The observation and evaluation of actual software use is a means to make this hypothesis explicit and to test it. In the case of invalidation, feedback is gained and a situation can be designed which is a better approximation.

This Section of the book is devoted to two main functionalities in problem-solving: planning and understanding. Actual computer aids to problem-solving have to support these functionalities in order to reach efficiency. As it has been shown (Wilensky, 1983), these two problem-solving attributes, require very similar processes.

Planning is the use or generation of representations capable of guiding the problem-solving activity: representation of procedure, state or evolution. Understanding is the use or generation of representational structures capable of matching incoming information. In both cases, elaboration and use of schematic representations is needed to manage complexity. In the planning context, this schematization is supported by the abstract problem space concept (Newell and Simon, 1972), while in understanding the concepts used are various: schema, plan, or MOP (Schank, 1980).

This introductory chapter outlines a general framework in which problem-solving, planning, and understanding, are considered in relation to computer support (a more extensive presentation can be found in Hoc, 1988a). Some elements of this framework and some needs for computer support are discussed in more detail in the chapters of this Section within quite specific problem-solving situations: text production (Bisseret), process control (Goodstein), and program understanding (Détienne).

At first, problem-solving situations are defined in terms of level of control of activity, stressing the link between problem-solving strategy and understanding. Then a wide point of view on planning is proposed which shares common attributes with understanding. Finally, some crucial issues about computer aids to planning and understanding are set.

2   PROBLEM-SOLVING AND UNDERSTANDING

Roughly, problem-solving can be contrasted with procedure execution. This opposition bears on a hierarchy of levels of control of the cognitive processes:

— from the highest level of very general domain-independent problem-solving strategies or heuristics, using wider knowledge than that which is strictly relevant in the domain, very often used by beginners;
— through domain-dependent strategies which characterize experts;
— to diverse stages in the process of automatizing action, until, at the lowest level, cognitive representation becomes useless and parallel processing possible.

This hierarchy may be translated immediately into terms of the evolution of activity, from learning to long practice. But, while it is clear that the lowest levels of control are only reached by experts, so that beginners fall into the highest levels, it would be too simple to class experts only into the lowest levels. Certain kinds of expertise combine automatisms and problem-solving activities, and it would be more true to say that expertise makes a wider repertoire of levels available to the operator.

In the design of computer-aided work stations, the best practice is surely not to bury the operator in automatisms or well-known procedures. That is certainly the case in process control, where the designer cannot foresee every possible event (Goodstein, this Section) and therefore cannot plan corresponding valid procedures to be taught. Very often the 'conditions of validity' of procedures are no more than implicit hypotheses, to be tested during the procedure execution; and the operator is confronted with problem-solving situations when hypotheses are invalidated.

The difficulty in analysing problem-solving strategies is the human ability to perform cognitive activity at different levels of control at the same time. The classification proposed by Rasmussen (1986: the most recent presentation)—skill-based, rule-based, and knowledge-based behaviour—is very useful in approximating these levels, but the entire activity leading to a goal can seldom be assigned as a whole to one level. Expert strategies articulate automatic schema triggering as well as constructive plan generation, routine application of well-known procedure, and procedure elaboration. This is why an activity is considered as problem-solving in relation to the highest level of control it requires to reach the goal (Bisseret, 1984).

It is useful to consider a problem as a subjective construct, relative to a particular cognitive system (biological or artificial). What is straightforwardly accessible is a 'task', which can have an explicit expression, for example in instructions (a prescribed task). A 'problem statement' is no more than a task expression. This wording is used in contexts where the observer guesses that the subject to whom the statement is given would have to develop a problem-solving strategy instead of executing a routine procedure to reach the goal.

Here a task is considered as a goal, with which some conditions of attainment are given (Leplat and Hoc, 1983; Hoc, 1988a). The goal can be more or less ill-defined: minimally, the goal is to leave the current state and try to reach a better one. The conditions can be seen as either constraints or aids:

— *constraints upon operations*. Only a certain kind of operation (available on a specific 'device') can be allowed, for example the operations coded by a programming language.
— *constraints upon intermediary states*. Only certain states are allowed, for example for safety concerns.
— *constraints upon procedure*. Only certain types of procedure are allowed, for example for economic reasons (in computer programming, procedures leading to reading the same file several times would be seen as nonoptimal).

So, a problem is defined as a representation (mental model) of a task, elaborated by a human subject (a cognitive system, in general), who does not have immediately an allowable procedure to reach the goal. Then the subject has to construct a procedure to solve a problem. This construction activity corresponds to a strategy which includes some kind of meta-operation independent of the task domain: for example, comparing the effects of two task-domain operations in order to choose the most promising, in relation to goal attainment. From this point of view, a problem is dependent on a subject. Hence, the same task can be a problem of one type for one subject, and for another subject it can be a problem of another type, or even not a problem at all.

Thus problem-solving involves two interrelated components:

— *problem understanding* (the construction of a coherent representation of the task to be done;
— *procedure searching* (the implementation of a strategy to find or construct a procedure).

Understanding processes are crucial in problem-solving, since the representations to which they lead correspond to the conditions of implementation of the strategies which are triggered. Very often problem-solving involves changes in representation which cause the subject to adopt successively diverse strategies before reaching the goal (Richard, 1984). Then problem understanding and procedure searching are two interrelated components of problem-solving: their implementations are rarely serial.

The way in which a problem is understood at a given time defines the type of the problem at this time. After Newell and Simon (1972), psychologists have been accustomed to formalize problems in terms of state transformation space. Nevertheless, from the analysis of problem-solving protocols, it appears that this kind of problem representation cannot always be inferred. Other kinds of representation are used which lead one to adopt a wider view of what can be a problem.

State transformation problems
It would be convenient to restrict this category to those problems for which the subject's representation (or understanding) of the task is actually a search for a path through a space of states linked by operators or commands (typically in using a pocket calculator or a text-editor for example). In most of the cases, programming problems are of this kind (see Détienne, this Section). Nevertheless, this kind of problem can be extended to continuous transformation cases where propagation through a causal network takes place instead of discrete step-by-step transformations from state to state.

Structure or rule induction problems
Here the task is represented as a search for relations between elements in order to discover a coherent representation useful for action (for example, diagnosis of system failures: see Goodstein, this Section).

Design problems
The task is represented as a search for a detailed representation of a goal satisfying a set of constraints (for example, architectural design, but also text production: see Bisseret, this Section).

3 PLANNING

It seems useful to consider planning as an interactive combination of two complementary mechanisms: the construction of plans and the use of plans. In fact, a new plan is often built from old ones, and in the course of using an old plan, a new plan can be created. Although a plan is very often understood in the limited sense of an action plan, it can be more generally considered as a representation capable of guiding activity. In this Section, Bisseret shows how a plan of the text, considered as a structured representation of ideas, can guide the text production process. A plan is schematic or hierarchically organized. So, planning implies the definition of abstract spaces, where some details are ignored. An important question concerns what kind of details are ignored.

Planning is more typically encountered in problem-solving activities. The chapters of this Section show different examples of plans used in solving different kinds of problems:

— In *text production*, a plan can be an organization of ideas, as opposed to definite expressions in text (Bisseret).

— In *process control*, a plan can be a relational structure of abstract functions implemented in the equipment or plant, which is useful in diagnosis (Goodstein).

— In *programming*, plans are formalized as schemata coming from different knowledge domains (Détienne).

Two main aspects of planning relevant to computer system design are stressed in this Section of the book: schematization and anticipation.

3.1 Schematization

Planning develops within abstract spaces. Rasmussen (1986: the most recent presentation) defines two orthogonal dimensions for the description of this hierarchical organization, illustrated in Goodstein's chapter. I propose a slight reformulation of these dimensions.

3.1.1 Refinement hierarchy

Rasmussen names this the 'whole-parts' or 'decomposition' hierarchy. Following this dimension a plan is refined by decomposing a whole in its parts. Conversely, plan abstraction on this dimension corresponds to what Piaget names a 'simple abstraction', the implementation of which bears on generalization. From a lower space to a higher one, details which are not relevant at a certain level of analysis are abstracted, to make clear the structure of the whole. But the representations used in the two spaces are of the same kind: the same kinds of variables, relations, and properties. This hierarchy is very often used and studied in human–computer interaction.

3.1.2 Implementation hierarchy

Rasmussen names this the 'abstraction' or 'means-end' hierarchy. Within this dimension, a plan is implemented, going from function or principle to its realization. Plan abstraction corresponds to the 'reflective abstraction' in Piaget's terms, enabling one to be aware of the control structure of a procedure and to access to the reasons for the success. Now the abstraction concerns implementation constraints and makes clear reasons for the implementation by general functions. The higher space uses a new type of representation: variables, relations, and properties are different (for example, from electrical laws governing an electrical boiler to thermodynamic laws governing heat exchanges). This hierarchy is seldom used in human–computer interaction design, possibly because in the designer's representation of a user only knowledge is needed at the implementation level. Goodstein (this Section) shows that this conception of users is no longer valid when they are confronted with failures for whose diagnosis they need access to the links between function level and implementation level.

If planning needs the availability of abstract spaces in which schematization can take place, the movements in the hierarchies can go from schematic representations to detailed ones (top-down) or in the reverse direction (bottom-up). This is well illustrated by the Hayes-Roth and Hayes-Roth (1979) model of 'opportunistic planning' and stressed by Bisseret (this Section) in text production.

3.2 Anticipation

Considered as a schematic representation, a plan is a working hypothesis, to be tested and, sometimes, questioned. As a matter of fact, a plan is an anticipative structure from two points of view.

3.2.1 Temporal anticipation

As far as a plan of procedure is concerned, the definitive test of the hypothesis represented by the plan takes place at the moment of the actual execution. The execution feedback leads to the confirmation or invalidation of the hypothesis which corresponds to the temporal anticipation. This view of anticipation is especially relevant for state transformation problems.

3.2.2 Anticipation of details

In the course of refining or implementing plans in problem-solving, a plan is also an hypothesis concerning the relevance of a high-level structure for reaching the goal. But, at the moment of the construction or recalling of the plan, detailed information about the refinement or the implementation is not available to test definitely the hypothesis. Information will be supplied later on.

A crucial component in planning is plan evaluation which can be seen as the test of an anticipative hypothesis in relation to these two types of anticipation. In each case some kind of simulation can be executed to test the plan before getting the whole information. In the case of temporal anticipation, simulation of the execution can be performed. Computer programming and process control are typical examples where this form of simulation takes place in order to test whether the program, or the mental model of the process, is correct.

More often than not experts are able to represent some crucial details of a plan before reaching the abstract space level corresponding to these details. This ability has been described in expert programmers by Adelson et al. (1985). This anticipation of details is very useful in evaluating plans or choosing those which solve interactions between subproblems as accurately as possible.

Especially when plans are considered as schematic structures, they are relevant to understanding as well as planning. They permit the subject to match environmental data to hierarchical structures capable of giving meaning to situations. Sometimes these plans are just recalled (as in Détienne's experiments, reported in this Section), at other times they are constructed from the detailed information.

4 AIDS TO PLANNING AND UNDERSTANDING

Most of the software tools supposed to aid planning or understanding present the user with two kinds of difficulties.

When they cannot be used before detailed information is available
Computer aids in control rooms deal with parameters instead of high-level conceptual representations of the process. Text-editors do not help to organize ideas, but characters, words, or paragraphs (Bisseret, in this Section, shows that more appropriate software is being designed in this direction, but is not yet satisfactory). CAD software in architectural design interacts with the architect when metric constraints are introduced: at this moment the design process is almost brought to an end (Lebahar, 1983).

Sometimes the schematic representations used by the operator are well known, at other times they have to be discovered; for example, when they are not expressed in ordinary conditions. What are the schematic representations used by the architect in designing? Cognitive analysis of this activity is needed to make them explicit. Indeed they are expressed in drawings, but only partly. What are the basic structures of the organization of ideas in text production? Bisseret (this Section) proposes some of them, but much work is needed to describe them exhaustively. Discovering the basic schematic representations used in planning and understanding in each domain is a necessary condition to design efficient computer support in the domain.

When they force the user to adopt a too-rigid strategy
In discussing the implications of psychological knowledge about planning and understanding in the design of computer aids, the question of the degrees of freedom left by the software to the user should be stressed. Three aspects of this question will be now addressed.

4.1 Linking Bottom-up and Top-down Strategies

As has already been noted, it is not convenient to restrict planning mechanisms to top-down components only. Clearly planning combines:

— *top-down components*, in creating new plans out of old ones or detailing plans (these components are made explicit in the formulation of top-down computer programming methods for example);
— *bottom-up components*, in elaborating new plans or adapting old plans out of detailed information gathered in the situation.

Difficulties in learning and implementing top-down problem-solving methods in programming are good examples of the necessity to combine bottom-up and top-down components in planning (Hoc, 1988b). The same case can be seen in text production (Bisseret, this Section), or program understanding where schema-driven components are linked to data-driven components (Détienne, this Section).

To be efficient, computer aids to planning and understanding must permit the user to shift easily from top-down to bottom-up strategy and to alternately make use of the advantages of each kind of strategy.

4.2 Linking Different Representations of the Situation

Several knowledge domains are often implied in a situation. So parallel access to different points of view on the situation is needed. Goodstein (this Section) suggests the use of multiwindowed displays to combine these representations (from functions to implementations). In the same way, different kinds of knowledge are called for in program understanding (Détienne, this Section): problem domain, data and algorithm structures, program and variables plans, etc. Programming languages cannot explicitly code all these kinds of content: other expression means must be used to fill the gaps in programming language expression.

4.3 Linking Different Representation and Processing Levels

Transition from problem-solving to execution seldom, if ever, is sudden. Very often these two kinds of activities are interrelated. So computer systems have to deal with an activity which can develop at different levels almost at the same time. Anyway, the point is that an activity is not stable in this hierarchy and very often combines different levels of control. This is the reason for the differences between what Rasmussen (1986) terms knowledge-based, rule-based, and skill-based performance, in the design and evaluation of computer aids. Rasmussen stresses that the needs for computer aids are different at different levels of control. But these diverse kinds of aid must very often be available at the same time, when the activity combines different levels of control.

In designing aids to problem-solving, an adequate allocation of the components of the activity among the different levels of control appears to be crucial. While working by hand, the problem-solver uses automatisms in the realization of certain subtasks (for example, editing or searching tasks). These automatisms permit the subject to reduce mental load at the level of mental representation and allocate resources to the main problem-solving process. Computer systems must be designed to satisfy the same function:

— either by taking charge of these subtasks,
— or else by proposing direct access commands, without tedious travelling through menus, in performing these subtasks.

5 CONCLUSION

The aim of supporting problem-solving and understanding leads the designer to define human–computer interaction in a quite open-ended way. Not only schematic representation processing has to be supported, but also combining different kinds of strategy components. The main difficulty in designing such computer support is the access to representations and strategies which are not overt in the subject's behaviour. The following chapters show how psychological investigation can be useful in gaining knowledge of these features of planning and understanding in different task domains, in order to specify requirements for aid. Moreover, they show that among the variety of domains some invariants remain, which have been described in this introductory chapter and can guide the approach of other domains.

REFERENCES

Adelson, B., Littman, D., Ehrlich, K., Black, J. and Soloway, E. (1985). Novice-expert differences in software design. In: *Human–Computer Interaction—INTERACT '84*. B. Shackel (ed.) Amsterdam: North-Holland, pp. 473–8.

Bisseret, A. (1984). Expert-computer aided decision in supervisory control. In: *Proceedings of IFAC '84*, 2–6 July, Budapest.

Hayes-Roth, B. and Hayes-Roth, F. (1979). A cognitive model of planning. *Cognitive Science*, 3, 275–310.

Hoc, J-M. (1988a). *Cognitive Psychology of Planning*. London: Academic Press.

Hoc, J-M. (1988b). Towards effective computer aids to planning in computer programming. In: *Working with Computers: Theory Versus Outcome*. G. C. van der Veer, T. R. G. Green, J-M. Hoc, D. M. Murray (eds). London: Academic Press, pp. 215–50.

Lebahar, J. C. (1983). *Le Dessin d'Architecte*. Roquevaire: Editions Parenthèses.

Leplat, J. and Hoc, J. M. (1983). Tâche et activité dans l'analyse psychologique des situations. *Cahiers de Psychologie Cognitive*, 3, 49–63.

Newell, A. and Simon, H. A. (1972). *Human Problem Solving*. Englewood Cliffs, NJ, USA: Prentice Hall.

Rasmussen, J. (1986). *Information Processing and Human–Machine Interaction: An approach to cognitive engineering*. New York: Elsevier.

Rasmussen, J. (1987). Cognitive engineering. In: *Human–Computer Interaction—INTERACT '87*. H. J. Bullinger and B. Shackel (eds). Amsterdam: North-Holland, pp. XXV–XXX.

Richard, J. F. (1984). La construction de la représentation du problème. *Psychologie Française*, 29, 226–30.

Schank, R. C. (1980). Language and understanding. *Cognitive Science*, 4, 243–84.

Wilensky, R. (1983). *Planning and Understanding. A computational approach to human reasoning*. London: Addison-Wesley.

TOWARDS COMPUTER-AIDED TEXT PRODUCTION

André BISSERET

INRIA, Rocquencourt, BP 105, 78150 Le Chesnay, France

1 INTRODUCTION

One of the phenomena characteristic of the design process, and it is one which appears particularly during the first phase of the process, is the changing of objectives. The process of preparing this chapter proved to be no exception: my first objective was reflected by the title given: 'The Psychology of Production'. But when I started to prepare the text, I modified this too-general objective and was thus forced to change the title! The text still concerns the production process and possible ways to assist it, but it is mainly focused on the specific field of text production.

Writing is, of course, an extremely widespread task, and yet at the same time, it is probably one of the least aided tasks. Being researchers, we are among the first to be interested by the possibility of computer-aided writing systems. In fact, as 'generating ideas' is part of the total writing process, we are almost always in the process of creating texts. But industry, towards which we direct our ergonomic work, is also a great producer of texts. Machine manufacturers, and in particular computer constructors, are also great producers of printed paper—the number of pages of manuals that they publish each year is enormous.

Computer-aided systems have already been introduced in other areas of design: CAO systems in technical design departments and architects' offices, and programming environments for computer programmers. But how far can we go towards the design of a workstation for writers, and how can cognitive ergonomics contribute to its specification? This is the problem that I shall discuss here, bringing together some results drawn not only from research into psychology and psycholinguistics, but also from informatics of aiding software and artificial intelligence.

1.1 Production and Reception

In research into cognitive activities, *production* is studied to a much lesser degree than *reception*, and this is particularly noticeable in the field of texts.

COGNITIVE ERGONOMICS:
UNDERSTANDING, LEARNING AND DESIGNING
HUMAN–COMPUTER INTERACTION

Whereas there has been a great deal of research into text comprehension, research into text production has been much less substantial and this is true as much in psychology as in artificial intelligence. In the case of psychology, Gould (1980) explains this as a consequence of a greater methodological difficulty: experimental paradigms based on comprehension tasks can be applied more easily than those based on production tasks.

Perhaps it has also been too readily assumed, at a theoretical level, that if the comprehension activity could be explained, then an explanation of the production activity would follow automatically. In fact, production is quite different from comprehension, even if they have some cognitive processes in common, and more research should be undertaken into the production process.

1.2   The Top-Down Assumption

The production process has been, and is still considered to be, based on a top-down model. This belief is shared by laymen and specialists in the organization of work. Very often it is represented as a sequence of phases where the output of one phase becomes the input for the next. Here is one example, chosen from a good number of others:

— preliminary study
— detailed study
— technical study
— implementation
— test
— utilization
— maintenance

More generally, in this perspective, designing and producing something mean defining an objective as a starting point, and breaking this objective down into a series of subobjectives. These, in turn, are broken down into sub-subobjectives, and so on until a sequence of elementary actions is designed that will achieve the product.

In the field of text production, the same schema is often admitted. Nobody has forgotten the wise advice of teachers: 'Start by making a plan, and everything will fall into place'. But how can a plan be made, and how can it be filled in? As far as I am concerned, when preparing this text, I kept on finding that once the plan has been formed, it remained rather unstable, and moreover that writing different parts of the text is a stressful activity, full of shifts and movements, in ideas, on paper, on the screen and even . . . up and down the corridor!

My plan is first to start with some important results drawn from works on planning models. Then I will review briefly research carried out on different human design activities, in order to focus finally and in more detail, on the writing process itself and on the problem of its assistance by informatics.

2 RESEARCH ON PLANNING

2.1 Hierarchical Model

Concerning planning models, it is true that an important line of research in artificial intelligence seems at a first sight to strengthen the common idea. This line has developed models of planning activity which are based on a completely top-down approach. These are the 'hierarchical planners' the best-known example of which is NOAH (for Nets of Action Hierarchies) produced by Sacerdoti (1977).

This type of plan takes the form of a hierarchical tree. The root represents the general goal of the activity, and the leaves represent the sequence of concrete actions that enable the goal to be reached. Each level is a complete representation of the plan but with a decreasing degree of abstraction from the root to the leaves.

However, the process of constructing this sequence of actions is carried out in a purely top-down way. The planner begins by breaking the general goal into a set of subgoals which is the second level. These subgoals are in turn refined in a third level, etc. But each time a new level is created, a 'criticism' is executed, in order to solve possible contradictions, interactions and/or redundancies. This technique, known as the 'least commitment' technique, was introduced as a powerful means of minimizing 'U-turns', goal modifications and/or changes in the order of the goals.

It is understandable that this type of model satisfies the specialist in artificial intelligence as far as they do not attempt to simulate human processes nor to interact with a user in a planning task. But as soon as we try to design tools for interactive aid, we come up against the problem of understanding, in order to take them into account, the differences between the logico-mathematically powerful representations used by program designers to maximize the power of their programs, and the various heuristic representations that people use in ordinary life, in order to adapt rapidly to different situations (see Bisseret, 1983).

2.2 Opportunistic Plan

In such a psychological perspective, Hayes-Roth and Hayes-Roth (1979) carried out an interesting analysis of verbal protocols from subjects who had the task of planning a set of errands, to be done in one day, in a town. The model they built from this was then implemented on a computer. This model is more comprehensive than the strictly hierarchical model: it does not contradict it but it includes it as a special case of human planning. This is the case when the task inherently presents a strongly hierarchical structure and/ or when the subjects are experts in the task. They may then use ready-made plans that they already have at their disposal and that they need only to invoke and adapt to the particular situation.

In the other cases, Hayes-Roth and Hayes-Roth clearly show the importance of the bottom-up process in the subjects' planning activity. They speak of opportunistic planning. Different levels of abstraction remain an important characteristic of this type of planning, but the subjects do not construct them according to a systematic refinement of the more abstract goal. The subjects work multidirectionally in the plane of abstraction. They frequently construct low-level subplans at the outset, which are relatively independent of each other and then combine them at a higher level of abstraction.

In particular, the subjects can build sequences of elementary actions which are not in keeping with a more abstract plan. Thus, they can work on a partial but seemingly promising aspect of the problem, without yet knowing how they will incorporate it into an overall solution. The authors note that in doing so, the subjects run the risk of producing incoherencies that will have to be corrected afterwards: an eventuality which purely top-down hierarchical planning avoids. But a specific advantage of the opportunistic strategy is that its bottom-up structure is an interesting source of innovation: low-level decisions and connected observations can lead to the discovery of new high-level plans.

3 EXPERIMENTAL RESEARCH INTO DESIGN

Now, I am going to show that the validity of the opportunistic model is supported by experimental research on different human design activities. For several years the IBM Behavioural Science Group has been studying design process in general to gain a greater understanding of the computer software design process (Malhotra et al., 1980; Thomas and Carroll, 1981).

From this group, Carroll and Rosson (1985) have recently published an interesting synthesis on the topic. They develop a convincing criticism of classical approaches, which they term 'analytical approaches'. They show that such approaches are more normative than experimental in that they represent the views of their authors 'about what design should be like but fail to seriously consider what design activity is like in fact'. Carroll and Rosson underline two faults in these analytical approaches: the first is that they consider design to be a top-down refinement of the problem in subproblems, and secondly, 'design appears to be a state instead of a process'. They show in particular how recent, well-known works on command languages, carried out by Moran (1981) and Reisner (1984), do not escape their criticism.

Carroll and Rosson, on the basis of precise empirical studies offer a radically different view of design activity which corroborates the results of Hayes-Roth and Hayes-Roth on opportunistic planning and which they summarize in the following way (Carroll and Rosson, 1985, p. 27):

— design is a process (it is not a state);
— design is non-hierarchical: neither strictly bottom-up nor strictly top-down;
— design is radically transformational, involving the development of partial and interim solutions which may ultimately play no role in the final design;
— lastly design intrinsically involves the discovery of new goals.

Visser (1987) is studying the activity of a technician in the design department of a firm where machine-tools are made. This technician is provided with the specifications of a machine and he is supposed to translate these specifications into the form of 'sequence-schemas' intended for the programmer of the automatic control system. The first protocol analyses show that the technician's activity is far from being a simple translation of the specifications. Not only does he fill in gaps and correct errors but he also takes decisions that put in doubt previous decisions and may even lead to modifying the design of physical characteristics of the machine.

In a fine book on the cognitive psychology of planning, Hoc (1987) emphasizes the interaction between top-down and bottom-up processes. The author carried out a great deal of research into the cognitive activity of software designers. Particularly in this field the normative model of refining a problem into subproblems tends to be put forward as being *the* solution. It is the panacea of 'structured programming'. Hoc's results clearly show that this method, however satisfying it might appear in itself, can hinder the subject's normal strategies.

Thus, studies on design in different fields, when they rely on precise empirical data, all show that the real design process as carried out by a human subject does not correspond to the ideal image sought by many people. In particular, designers of software for computer-aided tasks, generally, are too ready to believe that the activity of the future user is necessarily a 'beautiful tree'. It is true that the product itself can generally be represented by a nice hierarchical plan, although it is sometimes arbitrary (Tazi and Virbel, 1985a); but in any case, the plan of the product itself is one thing, the plan of the activity of production is quite another.

4 THE HAYES AND FLOWER WRITING-PROCESS MODEL

Let me now focus back on the text production process itself. I will first present the main characteristics of the writing model which has been proposed by Hayes and Flower (1980) and then I will review in more detail different subprocesses and the possibilities of their computer assistance. Hayes and Flower are the first authors to have studied the writing activity by means of the method of 'protocol analysis'. Proceeding this way they had the opportunity to analyse not only the product but mainly the process of its production.

The main characteristics of the model for our purpose are the following: it identifies the process as composed of different subprocesses and shows the organization of these subprocesses. These are:

— the process of planning which is composed of the processes of idea generating, idea organizing, and goal setting;
— the process of translating ideas into the form of language;
— and the editing process that involves correcting the text once it has been produced.

I would like to concentrate on the model's main dynamic characteristics that have been tested in a relatively precise way.

The authors show that these subprocesses are not organized as a strict sequence, from generating to editing. Actually, in a first phase, the more frequent subprocess is generating but is interrupted from time to time by editing. In a second phase, organizing is dominant but interrupted by generating and editing. In a third phase, translating dominates but is interrupted by generating and editing. Moreover, these interruptions are both frequent and widely distributed.

In the model it is the role of a 'monitor' to manage the interventions of the different subprocesses. The monitor makes it possible to simulate different writing strategies, corresponding to differences among individuals. An important aspect, which should be emphasized, concerns *the different types of physical traces* that characterize each subprocess. The generating process is characterized by the writing of notes which often consist of a single word or phrase and only seldom take the form of a complete sentence. The organizing process is also characterized by the writing of notes but here they take on a more complex form: they include physical marks such as numbering or indenting, that organize at least two ideas. Two general structures are represented this way:

— either a temporal structure ('I'll speak about design process in other domains and then about writing process')
— or/and a hierarchical structure (under '"writing aid" I'll speak about translating and planning').

The goal-setting process is characterized not by content, but by metacomments which evaluate the text. For example 'it would be more interesting to give examples from my own writing activity' or 'I'm needing a transition here'. The translating process is, of course, characterized by complete sentences that are intended to be included in the text. Finally, the editing process is characterized by all the corrections, both in the notes and in the text itself, and concerns all levels of language from spelling mistakes to pragmatic and rhetorical changes.

In summary, I underline two main points: Results on writing process are identical to those obtained in other areas of design. In fact writing is composed of several subprocesses which are not organized in a linear sequence of steps, but they interact throughout the writing activity. However, each of these subprocesses is clearly distinguishable. Each of them brings into play specific cognitive activities, which give rise to specific observable behaviours.

Taking advantage of this, I shall now review in more detail two of these subprocesses: the planning and the translating processes. For each, calling upon studies I found relevant, I shall pose the problem of possible specific computer aids, which could be integrated in a future writer's workstation.

5 COMPUTER-AIDED PLANNING

As main components of the planning process the Hayes and Flower model identifies the two subprocesses of generating and organizing. It should be said at the outset, that the term 'generating' is really something of an overexaggeration when it comes to the real process that is studied by research into writing. 'Retrieval' is also used and I find it more appropriate.

Even having taken this precaution, describing at a fundamental level of this retrieval process is still a research problem. It is dependent on progress in several areas of cognitive psychology, such as research into memory, learning but also problem-solving. That is a lot and I shall be satisfied here to simply underline several characteristics of the retrieval process, and more specifically its relation to the organizing process in so far as they are relevant to the problem of their computer assistance.

'From chaos to order' is the image that Brown and Newman (1985) used to sum up to the process of text production. But what is known of this chaos and the way it is ordered? Firstly, it is the subprocess that is the most time-consuming relative to the other components of the whole writing process. Gould (1980) in his study on letter writing, showed that planning accounted for two-thirds, on average, of the total time spent. An efficient computer-aiding system would thus be really worthwhile. A second point is the essential difference between the plan of the text and the plan of the activity. Hayes and Flower stress that the sequence of ideas retrieval is largely determined by a process of ideas association whereas the final text shows an obviously different plan which is plainly the result of the organizing process.

It is necessary thus to distinguish between the content plan and the process plan. The content plan is an abstraction of the complete text, whereas the process plan concerns the writer's strategies which is to say the goals that he/she sets and the means he/she uses to reach them. This process plan gives rise to metacomments that the writers make to themselves. The following is a significant example taken from one of the authors' protocols:

> 'I am really having a hard time getting started. Well, may be I'll just write a bunch of ideas down and may be try to connect them after'.

Such a conscious control of the process itself appears to be a specific characteristic of good writers. On the contrary, novices too quickly try to produce and manipulate text instead of explicitly manipulating ideas (see also Collins and Gentner, 1980).

Concerning possible ways to aid the retrieving and organizing process, it is clear that computer expertise cannot still be seriously considered. However, some kind of concrete support can be imagined which would allow the writers to materialize and manipulate their ideas on the screen. Such an hypothesis raises two questions. It is clear that texts consist of more elementary, well-defined units, such as words, sentences and paragraphs, that can be used as primitives for text editing systems. But what about idea as a unit? Is an idea an autonomous unit, a primitive that could be presented concretely in a computer-aided system for a direct manipulation of ideas? And secondly, what are the connections between ideas? Here again, is it possible and useful to provide the user with concrete presentations of them, and if so, to what level of detail?

5.1 Representing Ideas

It seems realistic to consider ideas as units. On the one hand, not only does the 'idea of an "isolable idea"' exist, but common observation, confirmed by studies on writers' protocols shows that actual behaviour corresponds with this. There are the one-word or single-phrase notes that are often distributed over separate sheets of paper.

Moreover, research into text comprehension, particularly the model from Kintsch and Van Dijk (1978), contributes to make precise the nature of ideas. According to this model, the sentences of a text are translated by the readers into sequences of semantic proposition (that is to say predicate-argument units). These propositions constitute the 'microstructure' of the text. Each proposition is interpreted not in an absolute way, but in relation to the others. As soon as the text reaches several dozen propositions, the readers develop a process of reducing the semantic information. They apply four rules, called 'macrorules'. Each using a different procedure, these macrorules allow the readers to eliminate the details of individual propositions and to construct chunks of signification, called 'macropropositions' which retain the global meaning of the text and which form a 'macrostructure' (see Van Dijk, 1977).

This rules-application process is recursive: as soon as the first level of macrostructure contains too many macropropositions, the readers (re)apply the rules to obtain macrostructures at a second level of generality, and so on. This is the process that allows us to remember for example, the content of a 500 page book. What is more is that, as concerns the problem of writing, at least in certain cases, the process is reversible. Taking a macroproposition, it is possible to rediscover its underlying micropropositions.

In the domain of computer programming an analogous distinction has been made between an expert's and a novice's representation of a program (which is a kind of text). Abelson (1984) shows that the novice's representation remains at the detailed level of 'how a program functions'. On the other hand, the expert's representation is at the level of 'what a program does': this is an abstract level (a macrostructure level) which no longer contains the details of the process.

Thus, we can conceive in a more precise way the fuzzy notion of 'idea' used in the studies on writing. An idea may well be considered as having a relative autonomy; ideas can be viewed as macropropositions with varying degrees of abstraction. In particular, each idea can be expanded at a more specific level.

5.2 Representing Relations

The second question concerns the possibility of concretely representing the semantic relations that the writer manipulates in order to organize ideas coherently and hierarchically. I think that this would not be a correct objective. Specific semantic relations are studied currently in different scientific fields such as knowledge representation, argumentation, explanation. But the results seem still too incomplete and widely dispersed. And above all, even if possible, it is unlikely that providing the writers with representations of specific semantic relations that they would have to manipulate would be an aid to them. Too many possible semantic relations would be a source of complexity and extra constraints rather than of efficient aid.

I believe that aiding systems, at least for a first generation, should keep to the only few semantic relations that are exhibited by the writers' observable behaviour, both in their final texts (see Tazi and Virbel, 1985b) and in the notes they produce (Hayes and Flower, 1980). Actually, there is a small set of very general relations between ideas which would be worth taking into account; these are mainly the sequence relation, the hierarchical relation and the relation of set membership.

5.3 Computer Aids Available for Planning

Is it possible to find tools that aid planning in this way? As far as I know, only plan *editors* are available on the market. But I do not believe that these editors are really aiding the planning process because they are based on a strictly hierarchical top-down model.

Here is an example of software which (in order not to advertise) I name 'Idea-lab' available on the 'Raincoat' computer. On the first page of the Idea-lab manual it says: 'the first program that offers the user the power of a personal computer to structure and organize his ideas'. According to the manual this software 'enables the user to organize his loose ideas by forming hierarchies and arranging them by topics'. The principle is compared with that of Russian dolls in order to highlight the power that the user can expect from Idea-lab! Idea-lab is faithful to the top-down model to such an extent that the manual underlines that: 'it is not possible to shift a title more than one level to the right of the title that precedes it'. And: 'it is forbidden to shift a title to the left of the one that immediately follows it'! And, of course, you cannot start in another way than in writing the general title (the very first prompt is 'UNTITLED'). (Two 'devices' constrain you to start with the title: plan editors and . . . conferences organizing committees!)

In fact, I personally very much appreciate using Idea-lab on my Raincoat computer; but only for what it can help to do and that is editing plans that are already well designed. But it is claiming far too much to say that it aids the plan design process itself. However, some researchers have moved in this direction.

Nanard et al. (1984), for example, have designed a computer-aided system the explicit aim of which is to set documents containing texts and illustrations. But the principles upon which their tool is founded seem suitable for ideas as well. They use a box model. A document is a hierarchical tree made up of boxes. A box is either a terminal object or else a group of boxes. Nonetheless, they explicitly choose to allow 'a mixed strategy' which is 'sometimes top-down and sometimes bottom-up'. Thus, they offer the users two ways of manipulating the boxes:

— on the one hand they can create a hierarchical tree of boxes, in the same way as in a classic plan editor;
— but on the other hand they can also create boxes that are independent from each other and store them, temporarily, in a 'box stack'.

In this stack the users may regroup the boxes in a bottom-up way. So they can move boxes, or groups of boxes in both directions: either from the tree to the stack, or from the stack to the tree.

Brown and Newman (1985) published some information about two tools they are experimenting on. The general principle that guides their design is 'to reify the stages of thought in the authoring process by giving concrete form to the products of each stage'. With the first tool, their objective is explicitly helping the writers in their process of 'formulation, clarification, and structuring of ideas'. Here again the tool allows the users to create individualized notes on the screen. They can 'shrink' these notes in order to concentrate on the relationships between groups of notes. To do this they can draw boxes that group sets of ideas and arrows that represent the relationships between them.

The second tool is designed for what the authors consider as a later phase, in which the writers expand their ideas in pieces of text. The machine provides the possibility of representing these pieces of text in the form of cards between which links can be established. The tool makes it possible to describe these links explicitly. The authors believe this to be an interesting

possibility because 'typing of links enables a separation between the semantic content of an individual idea and its relationship with other ideas'.

One of the interesting possibilities of these tools may be that they could
help the writers to distinguish more clearly in their whole activity between
the process of manipulating ideas and the process of translating these ideas
into a textual form. It is, however, necessary to take into account the
continuous interaction between all the subprocesses, which was brought to
light by the research into writing.

It is true that ideas are generated faster than they can be coherently
expressed; this is why writers have to resort to a kind of shorthand in their
notes in order to reduce the constraints of linguistic problems. But it is also
true that only when writing out certain ideas do the writers discover exactly
what they want to say and how this ties in with other ideas. Often the
expansion of one idea leads to the creation of new ideas and reorganizations
of the plan. It therefore seems necessary, I believe, that aids for the different
subprocesses should not be presented by different tools. One single tool
would allow the users both to distinguish the subprocesses and to move
freely from one subprocess to another.

6 COMPUTER-AIDED TRANSLATING

I am focusing now on the translating process. This process allows an internal
representation of an idea or of a set of ideas to be embodied in a well-formed
sentence or group of sentences.

The autonomy of the translating process has been confirmed experimen-
tally. McCutchen (1986) asked children to write texts. She varied the grades
and the degree of knowledge in the topic (which was football). The results
show that knowledge of the content plays an important role: the children
produced a much more coherent text when dealing with a topic that they
knew well. However, keeping the level of topic knowledge equal there are
still important differences in the coherence of the texts, according to the
grades. The author concludes that 'as children become more linguistically
sophisticated, they acquire generalizable discourse and linguistic skills that
they can use even when, and perhaps *especially when*, their topic knowledge
is sparse'.

Moreover, it is now accepted that the internal representation of knowledge differs from its representation using language. For one thing, the
internal representation is more complete than its written or spoken expression. Writers rely, in fact, on the knowledge that the readers already have, in

order to tell them only the minimum they need. For another, the internal representation does not take the form of sentences. The units of internal representation are semantic propositions of the predicate-arguments type, that are grouped into larger units that take into account their relations: these larger units are schemas, knowledge frames, scripts or semantic networks.

This structural aspect of knowledge is being widely studied at present both in psychology and artificial intelligence. But what about the process aspect? Although this process has indeed an autonomous existence, there are practically no psychological studies, as far as I know, that attempt to describe it directly. On the other hand, a relatively important activity is, at present, being developed in artificial intelligence under the name of 'text generation'. It concerns the generation of sentences to form short texts, usually about one paragraph long. Knowledge structures, such as frames or schemas, are used as inputs for the generating system.

Some try, using natural language, to answer questions asked by database users (McKeown, 1985). Others attempt to describe events either in the style of an encyclopaedia (Thompson, 1984) or else in journalistic style (Danlos, 1984, 1985). Another example of domain is the description of photographs of natural scenes (Conklin and McDonald, 1982). These studies do not try to model the translating process of human subjects and therefore their authors do not attempt to test their psychological validity. But their great interest is that they result in programs which *do* produce texts and that, to achieve this, the authors are obliged to define clearly, and find solutions to, all the problems that this translating process poses.

Without being able to go into detail here, let us say that they have all defined and solved two main categories of problem. Firstly, the problems of the necessary *decisions about content*. For a given semantic representation, what information can (or must) remain implicit, and what information must be explicit in the text, and in what order should this information be presented?

For example, to solve these problems in a scene-description system, Conklin and McDonald (1982) carried out experiments, asking the subjects to rate the *salience* of the various items that made up the scenes represented in the photographs. In other experiments they asked the subjects to provide written descriptions of the same photographs. They found that the degree of salience is 'quite sensitive to changes in the size and centrality of objects in the scene'. They also notice a strong correlation between the order of the objects according to a decreasing salience and the order in which the objects are mentioned in the written descriptions. The authors use the results to design 'a program that generates descriptive paragraphs comparable to those produced by people'.

Secondly, the problems of *linguistic decisions*. Which words from the lexicon are to be selected? Which syntactic constructions are to be used? How is the text to be divided into sentences? The text generator designers use, in particular, what they name 'presentation strategies' that are domain-independent frames, each one corresponding to a specific linguistic objective such as: a 'define X' frame; a 'describe X' frame; a frame 'analogy of X to Y'.

A specially interesting perspective is given by Danlos (1985) in reaction against modular approaches that treat decisions about content and linguistic decisions separately. Danlos shows that 'decisions about lexical choice, determination of the order of the information, segmentation into sentences and choice of syntactic construction are all dependent on one another'. To account for this she introduces in her generator on the one hand what she calls a 'lexicon-grammar' which includes both phrases specific to the domain and their associated syntactical properties, and on the other hand a 'discourse-grammar' which integrates all the linguistic decisions. Such a discourse-grammar specifies the relation between meaning and form; it establishes a correspondence between one semantic relation (for example, the causal relation that the author treats in detail) and the list of structures that allow it to be expressed. A discourse-grammar 'provides all the combinations that are both formally realizable and semantically appropriate'.

The author states in conclusion that although the results are coherent texts, with a correct syntax and a pleasant style, the amount of work necessary to have such texts eventually produced is very large. Not only is the lexicon specific to the domain, but a discourse-grammar is specific to a given semantic relation. To construct such a grammar demands a 'deep linguistic study of the possibilities offered by the language to translate the semantic relation concerned'. Danlos states that this line of research, although potentially very productive, is necessarily very lengthy.

However, from the point of view of possible aid for the human translating process this is the only way I can imagine. It is the way of an expert system. Two cases could be considered: in one case the users would be able to indicate what they want to say by filling in slots in the schemas or frames specific to the domain. This would constitute the input to the text-generator which would then carry out the process of transforming these elements of information into a written text in correct language. If need be, for certain sentences for example, it could produce several possibilities from which the users could choose one. In the other case, the users would write their text and the expert-system would complete the revision process, proposing improvements (see Vaughan and McDonald, 1986).

Certainly this is not going to be possible for many texts for a long time to come; and unfortunately I don't believe that we shall be able to use such systems for our scientific texts in the near future. But I do not think it unreasonable to imagine that such systems may well exist before too long for certain restricted (and well confined) domains. A good example that is of particular interest to cognitive ergonomics is, or course, that of writing manuals for hardware and/or software.

7 CONCLUSION

Deliberately this discussion has not included text-editors. Because the editing process is currently the only process that is really computer-aided the objective was to look for possibilities to aid more comprehensively the whole process of text production. In such a perspective, it is not only a matter of adding new aids but also of allowing the writers the complex intermixing of the subprocesses that research on writing has highlighted. To make such an integrated tool and even for the only editor component the designers should be less narrowly confined to a restricted view of editing. The current view is still inspired too much by the task of the typist which consists in recopying a text already composed.

In their review on the 'Behavioral aspects of text editors', Embly and Nagy (1981) criticized the view of editing process as 'a transformation from an existing string of symbols known as the source file (which in the case of initial text entry, may be null) to a new string of symbols, known as the target file'. This has not changed a lot. To achieve this transformation the word editors offer a set of functions, which are practically always the same.

That which is different and what most of the studies try to improve, is the dialogue mode that is supplied to the users in order to put these functions into operation. The main criteria are usability and learnability. I do not deny the interest of this work: the recent innovation of the direct manipulation has been remarkable, and by the way it is a good example of co-operation between computer science and ergonomics. However, even if it is now far easier, it remains that the users have to command continuously what they want at a very low level. Research should be directed towards greater computer involvement.

Exemplary research on this is that of Tazi and Virbel (1985b) and Virbel (1985) whose project is to produce an intelligent text editing and formatting system in which human expertise may be represented. Their project aimed at allowing the writers to avoid specifying their intentions using commands to inform the system about the final representation of their text. To do this the authors undertook a preliminary study to analyse all the natural marks that writers spontaneously introduce when writing either on a typewriter or by hand. Their working hypothesis is that the physical characteristics of text are non-discursive traces of metatextual speech acts, such as introduce, entitle, enumerate, underline. The originality of this research is that, although it is focused on the editing subprocess, it takes into account its relation to the other subprocesses in text composition.

That the research which focuses on the man–machine dialogue is useful and successful should not lead to the neglect of necessary research on a more fundamental and integrated concept of assistance. It was the goal of this chapter to suggest that such research should not be centred primarily on the tool (or the product) characteristics but above all on the analysis and modelling of the cognitive activities of the (potentially aided) writers.

ACKNOWLEDGEMENTS

I am grateful to Pierre Falzon, Jean François Richard and Willemien Visser who criticized first drafts and gave me very useful advice, especially about the (product) plan. I also want to thank Marie Pierre Laborne who helped in the editing and formatting subprocesses.

REFERENCES

Abelson, B. (1984). When novices surpass experts: The difficulty of a task may increase with expertise. *Journal of Experimental Psychology: Learning, Memory and Cognition*, 10(3), 483–95.

Bisseret, A. (1983). Psychology for man computer cooperation in knowledge processing. In: *Information Processing '83*. R. E. A. Masson (ed.). Amsterdam: North-Holland.

Brown J. S. and Newman, S. E. (1985). Issues in cognitive and social ergonomics: from our house to Bauhaus. *Human–Computer Interaction '85*, Vol. 1, pp. 359–91.

Carroll, J. M. and Rosson, M. B. (1985). Usability specifications as a tool in iterative development. In: *Advances in Human–Computer Interaction*, Vol. 1. H. R. Hartson (ed.). Norwood, NJ, USA: Ablex.

Collins, A. and Gentner, D. (1980). A framework for a cognitive theory of writing. In: *Cognitive Processes in Writing*. L. W. Gregg and E. R. Steinberg (eds). Hillsdale NJ, USA: Erlbaum.

Conklin, E. J. and McDonald, D. D. (1982). Salience: the key to the selection problem in natural language generation. In: *Proceedings of the Twentieth Annual Meeting of the Association for Computational Linguistics*, 16–18 June 1982, Toronto, pp. 129–35.

Danlos, L. (1984). Conceptual and linguistic decisions in generation. In: *Proceedings of the Tenth International Conference on Computational Linguistics*, 2–6 June 1984, Stanford, CA, pp. 501–4.

Danlos, L. (1985). *Génération automatique de textes en langues naturelles*. Paris: Masson.

Embley, D. W. and Nagy, G. (1981). Bahavioral aspects of text editors. *ACM Computing Surveys*, 13(1), 33–70.

Gould, J. D. (1980). Experiments on composing letters: some facts, some myths, and some observations. In: *Cognitive Processes in Writing*. L. W. Gregg and E. R. Steinberg (eds). Hillsdale, NJ, USA: Erlbaum.

Hayes, J. R. and Flower L. S. (1980). Identifying the organization of writing processes. In: *Cognitive Processes in Writing*. L. W. Gregg and E. R. Steinberg (eds). Hillsdale, NJ, USA: Erlbaum.

Hayes-Roth, B. and Hayes-Roth, F. (1979). A cognitive model of planning. *Cognitive Science*, 3, 275–310.

Hoc, J. M. (1987). *Psychologie Cognitive de la Planification*. France: Presses Universitaires de Grenoble.

Kintsch, W. and Van Dijk, T. A. (1978). Toward a model of text comprehension and production. *Psychological Review*, 85(5), 363–94.

Malhotra, A., Thomas, J. C., Carroll, J. M. and Miller, L. A. (1980). Cognitive processes in design. *International Journal of Man–Machine Studies*, 12, 119–40.

McCutchen, D. (1986). Domain knowledge and linguistic knowledge in the development of writing ability. *Journal of Memory and Language*, 25, 431–44.

McKeown, K. R. (1985). Discourse strategies for generating natural-language text. *Artificial Intelligence*, 27, 1–41.

Moran, T. P. (1981). The command language grammar: A representation for the user interface of interactive computer systems. *International Journal of Man–Machine Studies*, 15, 3–50.

Nanard, M., Nanard, J. and Falgueirettes, J. (1984). *Top down or Bottom up Approach for Document Structuration*. Centre de Recherche en Informatique de Montpellier, Research Report No. 15.

Reisner, P. (1984). Formal grammar as a tool for analysing ease of use: some fundamental concepts. In: *Human Factors in Computing Systems*. J. C. Thomas and M. L. Schneider (eds). Norwood, NJ, USA: Ablex.

Sacerdoti, E. D. (1977). *A Structure for Plans and Behavior*. Amsterdam: Elsevier.

Tazi, S. and Virbel, J. (1985a). *The Limits of the Hierarchical Model for Textual Representation*. Research Report LSI, Université Paul Sabatier, Toulouse, 1984.

Tazi, S. and Virbel, J. (1985b). Formal representation of textual structures for an intelligent text-editing system. In: *Natural Language Understanding and Logic Programming*. V. Dahl and P. Saint-Dizier (eds). Amsterdam: Elsevier, pp. 191–205.

Thomas, J. C. and Carroll, J. M. (1981). Human factors in communication. *IBM Systems Journal*, 20(2), 237–263.

Thompson, C. W. (1984). Object-oriented text generation. In: *Proceedings of the First Conference on Artificial Intelligence Application*, 5–7 Dec. 1984, Denver, Colorado, pp. 524–9. IEEE Computer Society Press.

Van Dijk, T. A. (1977). Semantic macro-structures and knowledge frames in discourse comprehension. In: *Cognitive Processes in Comprehension*. M. A. Just and P. A. Carpenter (eds). Hillsdale, NJ, USA: Erlbaum.

Vaughan, M. M. and McDonald, D. D. (1986). A model of revision in natural language generation. In: *Proceedings of the Twenty-fourth Annual Meeting of the Association for Computational Linguistics*, 10–13 July 1986, New York.

Virbel, J. (1985). Représentation et utilisation de connaissances textuelles. In: *Actes du colloque Cognitiva 85*, Paris, Cesta, 1985, pp. 879–83.

Visser, W. (1987). Abandon d'un plan hiérarchique dans une activité de conception. *Actes du colloque Cognitiva 87*, Paris, Cesta, 1987.

COMPUTER AIDS FOR DECISION-MAKING IN PROCESS CONTROL

L. P. GOODSTEIN

Risø National Laboratory, DK 4000 Roskilde, Denmark
(Present address: Ibsgarden 196, DK4000 Roskilde, Denmark)

1 INTRODUCTION

The introduction of computers in process control has greatly influenced the possibilities for the treatment of information in the control rooms of industrial plants—plants which themselves are becoming more centralized and complex with greater potential for serious consequences in the event of (human) errors. In addition, computers are being introduced in the design phase of these plants with the result that databases are being established which will also be of great value in operational planning and management.

This advanced technology has also had the effect that industrial plants are becoming more automated and do not rely on human intervention for the control of normally planned and predicted situations. However, the operating crew still has to cope with all the tasks which are badly structured, as well as the unforeseen events and disturbances which can arise. In this way, the role of the operator has shifted from manual controller to systems manager, problem-solver and decision-maker; hence the term *supervisory controller*.

However, in reality, the operating crew is part of a decision making team and, in accordance with the functional allocations of the designers, plays certain assigned roles in dealing with the process. Use of the word *team* reflects the fact that the supervisory control of such complex systems has actually to be a co-operative effort within a group consisting of the designers, the automatic (computer-based) control system and the operating staff. This three-way arrangement arises because the decisions of the designers are embedded in the automatic system as well as the training of the operators. Thus, the prerequisite for successful co-operation is that the designer ensures that the computer and the operators can work together in a positive way by taking advantage of their different and complementary information-processing abilities and their different knowledge about the

COGNITIVE ERGONOMICS:
UNDERSTANDING, LEARNING AND DESIGNING
HUMAN–COMPUTER INTERACTION

system, the environment, the goals, etc. This probably means that the framework in which this co-operation takes place should involve a dynamic allocation of decision functions between the 'partners' with appropriate feedback and communication facilities between them.

2 DECISION MAKING FRAMEWORK

Figure 1 (adapted from Rasmussen, 1976) describes a framework which encompasses the various types of information processing sequences that characterize a decision-maker's (human or computer) activities in dealing with a problem. These include the completely rational approach where the decision-maker 'climbs up the ladder' on the left-hand side while observing, making an identification of state, interpreting the implications and prioritizing goals. Thereafter, the decision-maker 'climbs down the ladder' on the right-hand side in connection with planning and carrying out the appropriate set of actions to achieve or reach the chosen target state.

The diagram also indicates alternative paths from the initiation of a decision-making activity to its conclusion. For example, in situations which seem to be familiar to the decision-maker, shortcut paths can exist in the form of a large number of association rules, e.g. IF xx, DO yy. It is these which form the basis for the veteran or experienced decision-maker's behaviour in dealing with the object system and often comprise the 'shallow expert knowledge' which is acquired for building current expert systems.

Thus, while in the real world the rational type of decision-making behaviour is probably mostly restricted to novices feeling their way, much can be said for trying to force experienced users to resort more to this type of response through a suitably designed display and controls interface. The reason for this is of course that the repertoire of quick and effective responses is not all-inclusive for every possible situation. Thus, in their interaction with the system, operators first have to be made to realize the inherent risk in a hasty response and thereafter be supported in more of a knowledge-based decision making sequence in order to solve their current problem.

3 HIERARCHICAL DECISION SPACE

Another important consideration in decision support is the decision context, i.e. the representation of the problem space. Rasmussen (1984a) has dealt with this in detail in his description of the different levels of abstraction and

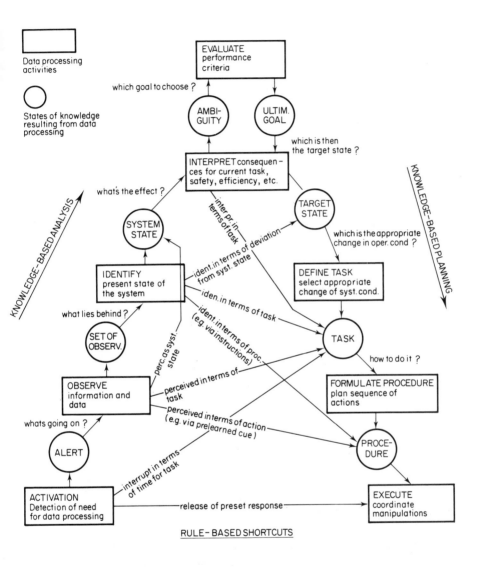

Figure 1. The sequences of information processes involved in a control decision. Rational, causal reasoning connects the 'states of knowledge' in the basic sequence. Stereotyped processes can bypass intermediate stages. (Adapted from Rasmussen, 1976.)

decomposition which a human may use to cope with the complexity of a technical system, depending upon the situation and the phase of a decision task.

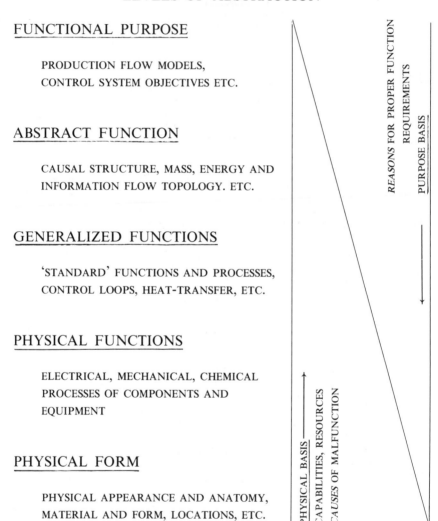

Figure 2. The system properties considered in man–machine interaction can be described at various levels of abstraction, representing the physical implementation and functional purpose in varying degrees.

In the abstraction, or means-end, hierarchy (see Figure 2), the low levels of abstraction are related to the available set of physical equipment which can be used to serve several different purposes. Models at higher levels of abstraction are closely related to specific purposes, each of which can be served by different physical arrangements. This hierarchy is therefore useful for a systematic representation of the many-to-many mappings in the purpose/function/equipment relationships which represent the context of (and are a necessary precondition for) supervisory decision making.

When considering a control task at any level of the hierarchy, information about the proper function, target states and answers to the question WHY is obtained from the level above, while information about present limitations and available resources, i.e. answers to the question HOW, can be got from the level below.

In the present context, we are particularly interested in those human functions in man–machine systems which are related to corrections of the effects of faults and other disturbances. States can only be defined as disturbed or faulty with reference to the planned or intended functions and purposes. Causes of improper functions depend on changes in the physical world. Thus they are propagating (and are explicable) *bottom-up* in the hierarchy. In contrast, reasons for proper functions are derived *top-down* in the hierarchy from the functional purpose. During plant operation, the task of the supervisory controller (man and/or machine) will be to ensure by means of proper actions on the system that the actual state of the system matches the target state specified for the intended mode of operation.

This task can be formulated at any level in the means-ends hierarchy. During plant start-up, for instance, the task moves bottom-up through the hierarchy. To have an orderly synthesis of the overall plant function during start-up, it is necessary to establish a number of autonomous functional units at one level before they can be connected to one function at the next (higher) level. This definition of functional units at several levels is likewise important for establishing orderly separation of functional units for shut-down and for reconfiguration during periods of malfunction.

During emergencies and major disturbances, an important supervisory control decision is the selection of the level in means-ends hierarchy at which to consider the control task. In general, highest priority will be related to the highest level of abstraction. First consider overall consequences for plant production and safety in order to judge whether the plant mode of operation should be switched to a safer state (e.g. stand-by or emergency shut-down). Next, consider whether the situation can be counteracted by reconfiguration of functions and physical resources. This is a judgement at a lower level representing functions and equipment. Finally, the root causes of the disturbance are sought to determine how the failure(s) should be corrected.

This involves the level of physical functioning of parts and components. Generally, this search for the physical disturbance is of lowest priority (not considering the role which knowledge about the physical cause may have for the understanding of the situation).

Thus, when a disturbance has been identified and the control task located at a certain level of abstraction, depending upon the perceived situation, the supervisory control task includes the determination of the appropriate target state derived top-down from the operation mode chosen together with an identification of the available functional resources and limits of capabilities, established bottom-up in the hierarchy.

4 SHARED DECISION MAKING

As stated earlier, decision making in supervisory control is a shared enterprise comprising the designer, the automatic computerized system and the operating crew. Since the designer may not be sure about his/her ability to foresee the necessary control responses for all possible disturbances (or because management and/or the regulators insist), there is need for a representative on site (the operator) who has to be able to take over in a competent way. The operator's supervisory control task is indeed in many respects a completion of the system design for the particular, perhaps infrequent, situation being dealt with. As a consequence, the operator will probably need information about the problem space underlying the design and the designer will have to communicate this kind of information to the operator. This can take place through the system itself; i.e. by means of the information gathered, processed and stored in the computer-based instrumentation and/or directly through training, manuals and instructions.

Sharing decision making among operator, designer and computer will result in a complex communication depending on the extent to which the designer wants his/her representatives on site to take an active part.

Criteria for the role allocation may be related to reliability requirements, to the resources and conceptual models available to support the processing by the different agents, or to the actual system state data accessible to them. Thus, the designer can choose to automate certain protective functions and thereby take over one of the supervisory subgoals; he/she can choose to pre-analyse certain phases of the decision sequence through use of analytical models and issue instructions to the operators; or, finally, a lack of data can lead the designer to transfer conceptual tools to the operators in the form of facilities for interactive decision making. These modes of role allocation will be considered in more detail in order to fomulate the communication requirements.

Thus, Figures 3a and b indicate how the basic framework of Figure 1 can be replicated to reflect how a designer might intend two typical situations to be allocated among the partners.

Figure 3a illustrates the situation where the designer will ensure a safe automatic shut-down of a plant in situations which are critical because of time constraints and/or potential consequences. (The various assigned roles are shown hatched.) The designer postulates sets of undesirable states (e.g. loss of control of core energy balance) and for each specifies:

— a set of process attributes/symptoms which are definitive for the particular undesired state;
— the desired target state (e.g. shut-down);
— the tasks to be performed to achieve the target state;
— the detailed switching sequences required.

The designer assigns to the computer the tasks of identifying the system state through the specified attribute set and, via a stored decision table which links symptoms to actions, of carrying out the automatic sequence.

After being informed via an early warning that automatic actions may occur, the operators' normal role is to continue with their supervisory task within the redefined operational envelope. In principle then, the goals of the computer and the operators are different for this situation. In practice, however, the designer will not be completely confident about the automatic system and the operators will be instructed to monitor that the automatic safety system functions properly AND TO INTERVENE IF IT DOESN'T. This completely alters the basis for the role allocation and, among other things, the information required by the operators regarding automatic shut-down as well as the control resources available to them.

Figure 3b is an example for situations which cannot be fully automated by the designer. Again the undesirable states and the associated symptom sets are identified. In addition, various redundant and/or diverse resources for coping with the postulated conditions are provided and procedures for the appropriate sequencing and control of these resources are generated—often labelled by the event which is to be dealt with. However, the actual diagnosis of the event is left to the operators—perhaps due to uncertainty or variability in the symptom patterns. Therefore, in this case, the designer/computer and the operators are sharing a task and will, in turn, take care of different subroutines in the decision process. The switching back and forth will take place at the standard nodes of the decision ladder and will involve the exchange of intermediate results. Operators are in charge of diagnosis (state identification) with help from the computer-based displays while goal priority and planning for action should come from a database supplied by

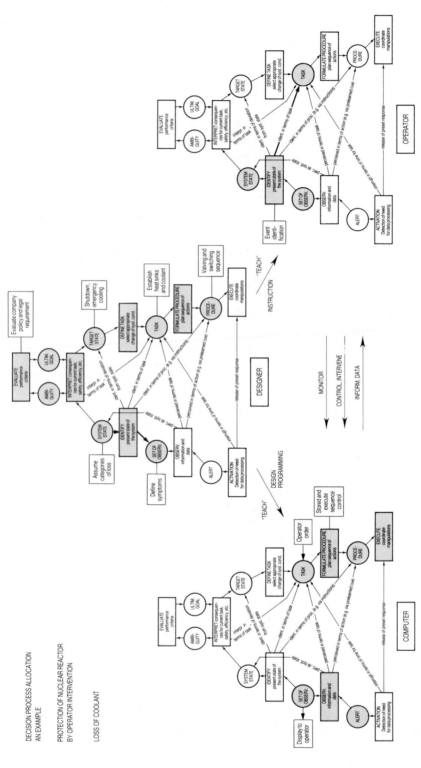

Figure 3a. The designer, the operator, and the process computer are all parts of a control decision. Their roles are shown (hatched) in an automatic safety shut-down, in which they act 'in parallel', each with a diagnostic task based on different strategies.

DECISION PROCESS ALLOCATION
AN EXAMPLE

AUTOMATIC SAFETY SHUTDOWN
OF NUCLEAR REACTOR

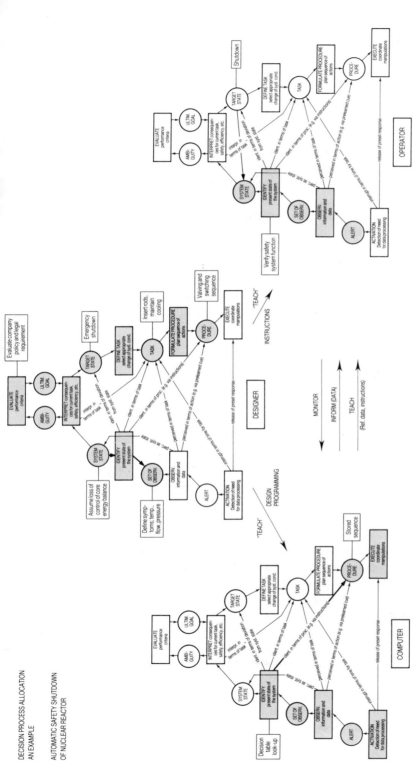

Figure 3b. Illustration of the roles of the designer, the operator, and a computer in a control decision which has not been fully automated, i.e. the operator and computer act 'in series'. The designer has automated a repertoire of protective sequences, but left the diagnosis to an operator. In addition to the decision functions, the designer, operator and computer support each other in different inform/teach/learn functions (Sheridan, 1982).

the designers. Subsequent requests for appropriate control actions will be carried out by the computer in response to orders by the operator.

There are again in this case some unclear elements in the planned allocation. If the operators are asked to treat the event-based procedures as irrevocable orders, difficulties will arise. Events are much more varied that the stereotyped categories which the designer can consider; in addition, the resources available at the moment will depend on maintenance schedules, errors during repair, etc.

On the other hand, if the instructions are intended to reflect recommended practice with room for variations in order to match specific occurrences, the operators will need further information about the designers' original analyses—especially the top-down specifications of intentions and goals and any other underlying assumptions and preconditions—in order to have the necessary references for judging the adequacy of the basic procedures as well as the acceptability of their own ideas for modifications to suit the current situation.

5 INTERACTIVE DECISION MAKING

The two examples have hopefully served to demonstrate the presence of a basic fork in the design philosophy road. Shall/can designers' analyses, heuristics and hypothetical foresight result in computer-stored decisions and guidance for all possible situations. Is this compatible with the policy of having well-trained, knowledgeable and experienced operators serving as final backup.

Alternatively, should one instead utilize modern information technology to place the designers' conceptual models and processing resources at the disposal of the operators who can then combine this design base with their up-to-date operational experience and knowledge in an integrated fashion and thus truly serve as the designers' extended arm in situations and disturbances which arise.

The argument in this chapter is in favour of the latter. In this case, the basic role allocation is that the operator and designer interact closely during the different phases of the decision sequence. Instead of communicating results of analyses based on hypothetical data, the designer brings conceptual tools into operation 'on site' through the computer. The computer has the capacity and accuracy necessary to test consistency of sensor data, to check correspondence with component characteristics and basic physical relationships (e.g. 'steam tables'), and to take account of mass and energy balances. In addition, analyses of the data collected during routine operations can be used for defining 'normal states' of the various functional levels for the different operating regimes which thereafter can serve as the references for locating disturbances in the functional topography. In this way, the computer will be able to interrelate actual states and target states up through the means-ends hierarchy. In fact, it thereby performs a diagnosis in terms of location of disturbances in the functional topologies without being dependent upon the designer's foresight. The results can be displayed as neutral statements of actual relationships up through the hierarchy without recommendations of actions or priorities.

It will be important to present results from the key nodes in the computer analysis and to select content and form of displays which will allow operators: (a) to make cross-checks with their own judgement which could be based on their empirical symptomatic data; and (b) to check hypotheses from such sources. The choice of priorities in goals will be left to operators and will depend on their perception of policies and available resources. In such a system, the mental load from having to integrate data will be diminished but at the price of the extra task of retrieving the proper displays or asking the proper questions to the database. Signals calling attention to the functional domain where changes have been identified by the computer (functional alarming; Goodstein, 1985) may be an efficient alternative to traditional alarm systems whose main intention is to give advice for action.

Judgement of resources is included in the activities in the downward planning leg of the decision sequence. Again, the communication from the designer should be based on a consistent engineering analysis of intended functions and on tools for on-line analyses of the available physical resources rather than hypothetical analyses of abnormal states. Frequently, several physically possible solutions may be available, and the choice will be dependent upon economic factors or maintenance status. For support of such prioritizing, an 'expert system' structure may be useful for the operating staff for recording overall plant experience. (Present 'expert system' technology probably will be more reliable in a planning task than for ranking choices in process plant diagnosis.)

This means that the computer and operators will share the planning function. A basic prerequisite for planning will be information on the operational state and availability of equipment and functions. Thus, computer support will depend on adequate access to actual configuration data, e.g. the actual state of the valving and switching. If this is available, the designer will be able to arrange displays of the possible configurations of equipment for various functions with indications of availability considering the maintenance states—the so-called 'success-paths' (Corcoran et al., 1981; Long, 1984). Procedural support for establishing higher-level functions on the basis of a choice of a proper success path can be available from a stored library of instructions which will be labelled neutrally by functions, rather than by events. For many such functions, automatic sequence control can be incorporated if necessary with input of actual conditioning information based on maintenance states from operators. This leaves the operator free to express the selected target states at different levels of integration, depending on conditions. Again, information from the computer is in neutral terms representing engineering analysis of defined technical functions and not hypothetical situations.

6 CONCLUSION

All of this has serious implications for the design itself, as well as the determination of the knowledge required by the operators. In practice, system design based on proven technology is largely an updating of previous designs with little specific attention being paid to a thorough task analysis and/or other special operator needs. However, a formal design and/or one based on new technology such as advanced computer-based techniques, requires a continuous iteration between considerations of purposes, functions and equipment in the means-end hierachy (Rasmussen and Lind, 1981)—also as the basis for defining the information to be made available to the operators to support their allocated supervisory roles. Indeed, as discussed in Rasmussen (1984b, 1985a,b), there is the need for a sytematic design and evaluation strategy which enables the designer to take into account the capabilities and limitations of the human as well as the computer so as to ensure that a well-balanced and error-tolerant design for meeting system requirements can be realized. In our minds, this amounts to carrying out a multiphase COGNITIVE TASK ANALYSIS which essentially is a mapping which must bridge the gap between, at the one end, engineering analyses of the target system (power plant, etc.) and, at the other, psychological descriptions of human performance. To be operational, the concepts used in the mapping also have to be consistent and compatible with, as well as expressible in terms used in the design of, computer and control systems.

The approach to computer support of supervisory decision making advocated here is to consider operators as being capable of taking on the authority and responsibility for the decisions required. Rather than continue a development where designers attempt to pre-analyse all abnormal situations and to store their advice in computers, they should instead try to make available to operators their conceptual tools and use the capacity of computers to perform on-line analyses of the available measured data.

Supervisory control should be based as far as possible on consistent engineering analysis. Heuristics and hypothetical foresight should be used to supplement such analyses and to guide the priority of choice, and not to replace on-site engineering analyses. However, real life is not black or white; thus future systems will probably have to combine all the approaches.

When an information system contains a mixture of factual information, heuristics and advice based on other people's foresight, credibility may be a problem. For the design of decision support systems, design criteria are necessary for enhancing users' acceptance of advice and establishing the preconditions for user understanding of explanations and justifications. A basis is needed for establishing and maintaining co-operative attitudes towards a computer intermediary.

REFERENCES

Corcoran, W. R. et al. (1981). The critical safety functions and plant operation. *Nuclear Technology*, 55(3), 690–712.

Goodstein, L. P. (1985). Functional alarming and information retrieval. *Risø-M-2511*.

Long, A. (1984). Computerized operator decision aids. *Nuclear Safety*, 25(4), 512–24.

Rasmussen, J. (1976). Outlines of a hybrid model of the process plant operator. In: *Monitoring Behaviour and Supervisory Control*. T. Sheridan and G. Johannsen (eds). New York: Plenum.

Rasmussen, J. and Lind, M. (1981). Coping with complexity. *Risø-M-2293*.

Rasmussen, J. (1984a). Strategies for state identification and diagnosis. In: *Advances in Man–Machine Systems Research*, Vol. 1, W. B. Rouse (ed.). New York: JAI Press, Greenwich, Co.

Rasmussen, J. (1984b). Human error data. Facts or fiction? *Proceedings of 4th Nordic Accident Seminar*. VVT. Symposium 55, ESPOO 1985, Finland. Also *Risø-M-2499*.

Rasmussen, J. (1985a). The role of hierarchical knowledge representation in decision making and system management. *IEEE Transactions on Systems, Man, and Cybernetics*, SMC-15(2).

Rasmussen, J. (1985b). Risk and information processing. *Risø-M-2518*.

Sheridan, T. B. (1982). Supervisory control: Problems, theory and experiment for application to human–computer interaction in undersea remote systems. *Dept. Mech. Eng. M.I.T. Tech. Rep.*, March.

PROGRAM UNDERSTANDING AND KNOWLEDGE ORGANIZATION: THE INFLUENCE OF ACQUIRED SCHEMATA

Françoise DÉTIENNE

INRIA, Rocquencourt, BP 105, 78150 Le Chesnay, Cedex, France

This study concerns the understanding of programs by experts in programming. Understanding a program, or more generally a text, necessitates constructing a representation of the program or the text. This elaboration involves different processes: decoding, interpretation of the linguistically coded information and activation of the knowledge stored in memory. The processes involved in the elaboration of a text representation are dependent on the organization of knowledge in memory and on the mode of access to this knowledge. So, the topic of program understanding raises the question of the processes involved in the construction of a circumstantial representation of a particular program with regard to the organization of the knowledge stored in memory.

An attempt to construct a model that accounts for the organization of the knowledge in a particular domain, has to take into account the expertise of the subject in the domain. Many studies (Adelson, 1981; Chi et al., 1981; McKeithen et al., 1981; Weiser and Shertz, 1983) gave prominence to the different organization of experts' and novices' knowledge in a domain. The expert in a domain not only has more knowledge than the novice, but, more importantly, he/she organizes that knowledge differently. In this study, we are only interested in the analysis of the knowledge organization of experts in programming.

1 THEORETICAL FRAMEWORK: THE SCHEMA THEORY

Our theoretical framework is based on the schema theory. This theory concerns the knowledge organization in memory and the role of that knowledge in text understanding (Schank and Abelson, 1977; Rumelhart, 1978, 1981; Anderson, 1981). A *schema* is a data structure for representing, in memory, generic concepts of a domain. It includes variables and constraints for the instantiation of the variables. A schema is instantiated when a particular configuration of values is bound to the variables, at a particular moment. These structures of knowledge account for the inferences realized in the processes of text understanding. While reading a text, a reader evaluates constantly hypotheses on the possible interpretation of that text. 'Readers are said to have understood the text when they are able to find a configuration of hypotheses (schemata) which offer a coherent account for the various aspects of the text' (Rumelhart, 1978). The processes involved in the elaboration of a representation are interactive, i.e. conceptually driven and data driven. By a bottom-up process, information extracted from the code activates schemata: this activation creates expectations on the instantiation of other slots of the activated schemata. The main process of understanding is the elaboration and evaluation of hypotheses (schemata). Some indices extracted from the code suggest possible interpretations by the activation of schemata; then, these interpretations are evaluated by information decoded from the text.

2 EXPERIMENTAL PARADIGM

With this theoretical framework as a reference, three experiments have been conducted. Two of these experiments have been presented in more detail elsewhere (Détienne, 1984, 1985). The experimental paradigm implied studying two aspects of experts' understanding:

— the process of representation construction was investigated using analyses of the subjects' activities during information gathering in a debugging task;
— the result of this activity, that is the representation stored in working memory, was investigated using a recall task.

This paradigm allows one to study the process of schemata activation during reading and the influence of the activated schemata on the form and content of the program representation.

During the reading phase, the subjects had to discover an unknown program and to evaluate its correctness. The programs were written in PASCAL. In a first experimental setting, we have recorded, during the reading phase, the verbalization of subjects, their written notes, the sequences in reading the different instructions. The study of the exploratory strategy of subjects shed light on the search of indices that activate their knowledge. To study the processes of elaboration and evaluation of hypotheses, another experimental setting was designed in which subjects had to verbalize: the program was presented one instruction at a time, and the subjects had to tell the information provided by the new instruction and the hypotheses that they could elaborate concerning the other instructions and the functions performed by the program. This experimental setting, which is not very natural, allows one nevertheless to find out the nature of the indices that activate, in general, the knowledge of the subjects, experts in programming. This information will allow the structure of the schematic knowledge possessed by experts to be defined. The verbal and behavioural protocols have been coded in the form of production rules: IF A THEN B. 'A' is a piece of information extracted from the code or the memory, 'B' is a piece of information allowing expectations on the program. These rules constitute a description of the activity of subjects. These rules have been categorized in general rules when different instantiations of the same rule have been observed. The regularities in the knowledge that are revealed by these rules have been translated in the form of a schematic organization that is supposed to be possessed by experts.

The recall task allows analysis of how the experts' knowledge affects the form and the content of the resulting representation. On this aspect, the schema theory makes different predictions: (a) distortions of the form of the program, (b) thematic intrusions, that is distortions of the content of the program.

The schema theory predicts distortions of the form of the recalled information, i.e. the effect of the process of abstraction. A subject does not keep in memory the surface form of the information presented but this information is abstracted in a more general representation. The information selected by the encoding process is reduced by the abstraction process. This process encodes the meaning of a particular piece of information but not the surface form, therefore, the lexical form of a concept or the syntactic form of a sentence may not be kept in memory.

The theory also predicts distortions of the content of the information, called 'thematic intrusions': they are observed when information typical of a schema is recalled while not included in the program text.

The results of these experiments give support to the hypothesis that the experts' knowledge is organized in schemata of different levels of abstraction. According to the goals of subjects, we have constituted three categories of rules that are descriptive of the activity of subjects: (1) rules of identification of functions, data structure, interaction between schemata; (2) rules of identification of algorithmic categories and particular algorithms (combination of elementary schemata); (3) rules of identification of elementary schemata. These rules allow description of the organization of the knowledge that is involved in program understanding. The experts possess schemata of different levels of abstraction: (1) schemata dependent on the task domain; (2) algorithmic schemata; (3) elementary schemata of programming. In these schemata, prototypical values could be associated with variables. These different issues are illustrated in the following text.

3 EXPERTS POSSESS SCHEMATA DEPENDENT ON THE TASK DOMAIN

Part of the knowledge possessed by experts is organized into schemata that are dependent on the task domain. During the reading phase and representation construction, an exploratory strategy was observed: in the first phase, the experts read the program rapidly by taking information on the declarative section, on the names of subprograms and on the body of the main program. It is possible to assume that the experts search for indices that inform them about the data structure and about the different processing phases. The analysis of the simultaneous verbalizations obtained in this reading phase shows that some indices of the code allow the experts to elaborate hypotheses about the task domain, the data structures and the functions performed in the program. Here are some examples described by rules of the form 'IF . . . THEN . . .'. The indices that the subjects extract from the code are on the left-hand side of the rule. The hypotheses elaborated by the subjects are on the right-hand side of the rule.

R
IF title of the program
THEN task domain
EX:
IF the title of the program is "stock management"
THEN the task domain is stock management

R
IF task domain
THEN data structure
EX:
IF the task domain = stock management
IF the type of variable1 = record with two fields
IF the type of variable2 = chain of characters
THEN the first field of the record = name of file
the second field of the record = descriptor of file
type of the first field = chain of characters

type of the second field = integer or real
first field = variable2

R
IF task domain
THEN performed functions
EX:
IF the task domain is stock management
THEN the performed functions are allocation (creation or insertion), destruction, search

The knowledge that allows elaboration of these expectation systems can be described under the form of schemata dependent on the task domain. The general structure of this schema includes slots on the data structure and on the possible functions in the particular task domain. The slots are the following:

task domain
data structure
functions

For the particular domain of stock management, the following values could be assigned to the slots:

task domain: stock management
data structure: record (name of file, descriptor of file)...
functions: allocation (creation or insertion), destruction, search

In one of our experimental settings, after the first phase of the exploratory strategy, the subjects had to perform a recall task. A thematic insertion was observed. The function of creation was not isolated in a subprogram and the subjects did not read any information concerning this function in the program. Nevertheless, a subject has reported this function as if it was a subprogram. The schematic knowledge dependent on the task domain has participated in the elaboration of the representation and in the process of recovering stored information at recall.

4 EXPERTS POSSESS ALGORITHMIC SCHEMATA

The analysis of verbal and behavioural protocols of subjects obtained during the reading phase, allowed elaboration rules for the identification of algorithmic categories and of particular algorithms (combination of elementary schemata). Here are some examples.

R
IF type of data structure (DS)
IF function
THEN algorithmic category for realizing function
EX:
IF type of DS = array of records without chaining structure (/with chaining)
IF functions = allocation, search, destruction
THEN algorithmic category = scanning (/chaining)
EX:
IF type of DS = array of characters
IF function = coding
THEN algorithmic category = haschcoding
EX:
IF type of DS = array of characters
IF function = reading
THEN algorithmic category = linear reading

R
IF ELEMENTARY SCHEMATA OF VARIABLE
IF function
THEN algorithmic category for realizing function
EX:
IF ELEMENTARY SCHEMA OF COUNTER
IF ELEMENTARY SCHEMA of READING VARIABLE
IF function = reading characters
THEN algorithmic category = linear reading
EX:
IF ELEMENTARY SCHEMA of COUNTER
IF function = search
THEN algorithmic category = linear search

R
IF algorithmic category
IF ELEMENTARY SCHEMATA of VARIABLE
IF ELEMENTARY SCHEMATA of Control structure
THEN combination of ELEMENTARY SCHEMATA

These examples allow description of the slots that could compose algorithmic schemata:

function
type of DS
elementary schemata of variable
elementary schemata of control structure

In a schema for linear search, the slots could take the following values:

function: search
type of DS: array
elementary schema of variable: counter
elementary schemata of control structure: iteration, test

5 EXPERTS POSSESS ELEMENTARY SCHEMATA OF PROGRAMMING

Some of the knowledge possessed by the experts is organized in schemata that are independent of the task domain. Several experimental observations support this idea.

During the reading activity and representation construction, the experts are able to diagnose some bugs without having understood what the program and the bugged subprogram perform: for example for the bugs of the initialization of variables used as counters. They have constructed a representation that is specific to a type of programming process and independent of the knowledge on the task performed by the program.

The analysis of the verbalizations obtained during reading illustrates the nature of the indices extracted from the code allowing the expert to make hypotheses on the semantic of the variables. Some indices allow the activation of expectation systems: the name of a variable and its type allow the subjects to expect particular forms of initialization, updating and a context. This happens whatever the knowledge the subjects have on the task domain. Here are some examples.

R
IF type of variable
IF name of variable
THEN initialization

THEN updating
THEN context
EX: (in the case of a variable, V, used as a counter)
IF there is a local variable named I, J or K,
of which the type is integer
THEN the variable is initialized at the value 0 . . . n
The variable is updated by an incrementation: $V := V + 1$
The variable appears in the context of an iteration
EX: (in the case of a variable used as a flag)
IF there is a global variable named Termine,
of which the type is boolean
THEN the variable is initialized at the value false
the variable is updated at the value true
the variable appears in the context of an iteration

These two examples allow description of the slots that could compose schemata of variables. The slots are of the following types:

type of variable
name of variable
initialization
update
context

The domain of the possible values associated to each of the slots is different for a counter and for a flag.

In a schema for a counter, the slots could take the following values:
type of variable: integer and local
name of variable: I, J, K . . .
initialization: $V := 0$ to n
update: $V := V + 1$
context: iteration

In a schema for a flag, the slots could take the following values:

type of variable: boolean and global . . .
name of variable: termine . . .
initialization: $V := $ false
update: $V := $ true
context: iteration

These constraints on the instantiation of the slots allow subjects to elaborate expectation systems during reading of a program: knowing the value associated with the slots 'type of variable' and 'name of variable', the experts are able to expect the values associated with the other slots of the schema.

Our hypothesis is supported by the form of the distortions observed in the recall protocols. Some distortions are made on the name of a variable used as a counter: for example, the subjects recall I instead of J. Then, the variables are memorized in a category, a schema for a counter, and the lexical form is not kept in memory. In the recall process, the subjects use another possible value of the slot 'name of variable'

Other observations suggest the existence of prototypical values in the slots of a schema. Each slot is associated with a set of possible values, as we have seen. These values have not the same status, some values being more representative of a slot for a schema than the others. When there is a prototypical value in a category, this value comes first in mind when the category is activated.

In a schema for a flag, the slot 'context' is an iteration that can take the values 'repeat' or 'while'. In the program used, this variable, V, appears in a 'repeat' iteration. During the reading phase, the value of the iteration expected by most of the subjects was 'while not V do'. In the recall protocols, a distortion was observed: a subject has recalled the instruction as 'while' instead of 'repeat'. This subject has reported the prototypical value instead of the adequate value. These observations give support to the hypothesis of prototypicality of some values associated with the slots of the schemata.

6 RELATIONSHIP BETWEEN SCHEMATA

The schemata represent knowledge at different levels of abstraction. The elementary schemata cannot be broken up into subschemata and can be matched against the code. The schemata dependent on the task domain represent abstract knowledge. The algorithmic schemata allow mapping between elementary schemata and schema dependent on the task domain.

Our observations support the hypothesis that some functions represented in schemata dependent on the task domain are related in memory to categories of algorithms. For example, the function of search could be related to many algorithms that could be organized, in memory, in categories and subcategories as algorithms of linear search or of direct access search. At a certain level of representation corresponding to a subcategory (for example linear search in a particular data structure), knowledge could be organized in a schema where slots are, on the one hand, elementary schemata, for example schemata of count variables, and, on the other hand, constraints for the combination of the elementary schemata, with some combinations being more prototypical than others. The elementary schemata could be defined as 'primitives'.

7 RELATIONSHIP WITH SOME MODELS OF PROGRAM UNDERSTANDING

These results support theoretical models of program understanding: the model developed by Brooks and the model developed by Soloway and Ehrlich.

We have observed that, very early, during the reading of a program, the experts search for indices that activate schematic knowledge dependent on the task domain. That knowledge allows the subjects to elaborate hypotheses on the data structure and on the possible functions of the program. This observation is consistent with Brooks' model (Brooks, 1983). According to that model, the process of understanding is based on the successive refinement of hypotheses about the knowledge domains and their relationship to the executive program. This theory asserts that the hypothesis generation and verification process begins with a primary hypothesis that is generated as soon as the programmer obtains any information about the task that the program performs: it will specify at least the global structure of the program in terms of inputs, outputs, major data structures and processing sequences.

Our observations suggest the existence of elementary schemata of programming. This notion is very close to Soloway and Ehrlich's concept called 'Programming Plan' (Soloway et al., 1982; Soloway and Ehrlich, 1983, 1984; Johnson and Soloway, 1985). Programming plans are program fragments that represent stereotypic action sequences in programming. There are two types of schemata: variable plans and control plans. Variable plans are plans which generate a result that is usually stored in a variable. This is very close to the elementary schemata that we have described for two types of variables, counter and flag.

8 PROSPECTS: SCHEMA ACTIVATION AND EVALUATION PROCESS

The present study has analysed some of the knowledge possessed by experts. This knowledge has been described in the form of schemata. Other studies will have to be conducted to complete this analysis: the hypothesis of schemata dependent on task domain should be evaluated in task domains other than stock management, the hierarchical relationship between schemata should be described with more accuracy, and the hypothesis concerning the schematic organization of experts' knowledge should be evaluated with languages other than PASCAL.

Whereas the processes of schema activation and schema instantiation have been well described by different studies in the past, there is a fundamental process that has not been studied before. That is the process of evaluation by which the subject evaluates the coherence of the constructed representation. We think that the evaluation process could be studied through the analysis of the failures to understand: when a subject does not understand, it means that he/she cannot integrate new information into the representation constructed before. The subject expresses a judgement of incoherence between new information and representation constructed before. We have already collected some observations on the failures to understand. The analysis makes clear that the subjects elaborate and evaluate hypotheses of external coherence and of internal coherence. The hypotheses on the external coherence between schemata are elaborated at the level of functional representation and depend on rules of interaction and causal relationship between performed functions. The hypotheses of internal coherence depend on constraints internal to schemata. In order to judge that some portion of the program is correct, the subject has to elaborate and validate hypotheses of coherence of these two types. Other studies will have to be conducted to test this hypothesis. These studies would give us information about debugging activity in which the evaluation process is most important.

REFERENCES

Adelson, B. (1981). Problem solving and the development of abstract categories in programming languages. *Memory and Cognition*, 9(4), 422–33.

Anderson, R. P. (1981). Psychological status of the script concept. *American Psychologist*, 36(7), 715–29.

Brooks, R. (1983). Toward a theory of the comprehension of computer programs. *International Journal of Man–Machine Studies*, 18(6), 543–54.

Chi, M. T. H., Feltovich, P. J. and Glaser, R. (1981). Categorization and representation of physics problems by experts and novices. *Cognitive Science*, 5, 121–52.

Détienne, F. (1984). Analyse exploratoire de l'activité de compréhension de programmes informatiques. *Proceedings AFCET, Séminaire "Approches quantitatives en Génie Logiciel"*, 7–8 June, Sophia Antipolis, France.

Détienne, F. (1985). Programming expertise and program understanding. *Ninth Congress of the International Ergonomics Association*, 2–6 September, Bournemouth, UK.

Détienne, F. (1988). Une application de la théorie des schémas à la comprehension de programmes. *Le Travail Humain* 51(4), 335–50.

Johnson, W. L. and Soloway, E. (1985) PROUST: Knowledge-based program understanding. *IEEE Transactions on Software Engineering*, 11(3), 267–75.

McKeithen, K. B., Reitman, S. R., Rueter, H. H. and Hirtle, S. C. (1981). Knowledge organization and skill differences in computer programs. *Cognitive Psychology*, 13, 307–25.

Rumelhart, D. E. (1978). *Schemata: the building blocks of cognition*. Center for Human Information Processing, University of California, San Diego, CHIP 79.

Rumelhart, D. E. (1981). *Understanding understanding*. Center for Human Information Processing, University of California, San Diego, CHIP 100.

Schank, R. and Abelson, R. (1977). Scripts, plans, goals and understanding. LEA.

Soloway, E., Ehrlich, K. and Bonar, J. (1982). Tapping into tacit programming knowledge. *Human Factors in Computer Systems*, 15–17, 52–7.

Soloway, E. and Ehrlich, K. (1983). Reading programs is like reading a story (well, almost). *Proceedings of the Fifth Annual Conference of the Cognitive Science Society*, 18–20 May, Rochester, NY.

Soloway, E. and Ehrlich, K. (1984). Empirical studies of programming knowledge. *IEEE Transactions on Software Engineering*, 5, 595–609.

Weiser, M. and Shertz, J. (1983). Programming problem representation in novice and expert programmers. *International Journal of Man–Machine Studies*, 19, 391–8.

INDEX

257